PRAISE FOR *Mine's Bigger*

"The man, his ego, and his boat are examined with insight and precision by David A. Kaplan in *Mine's Bigger*. . . . That *Mine's Bigger* is as much a biography of Perkins as it is the story of his boat isn't surprising. Kaplan is a fine reporter, mining sailing lore to give *Mine's Bigger* unexpected depth. Even jargony passages on the development of the *Falcon*'s sails come alive through Kaplan's understanding of nautical engineering and his evident passion for the sea. *Mine's Bigger* is a fine way to get to know Perkins and the story of how his big boat got built."

—*Forbes*

"An exhilarating account of how Tom Perkins, the man most responsible for bankrolling Silicon Valley, created 'the perfect yacht'—the biggest, fastest, riskiest, highest-tech, and most self-indulgent sailboat ever: the *Maltese Falcon*."

—*American Heritage*

"Engaging and revealing. . . . If it's your idea of fun is to admire the fabulously wealthy, brilliant, and charismatic person you will never be, you'll want to read *Mine's Bigger*. But if holding up the ludicrously self-involved for public examination makes you whimper with delight, you'll like it just as much. . . . Billed as the tale of his quest to build a mega-yacht to outdo all other mega-yachts, *Mine's Bigger* is really a biography of this fascinating, important, and frequently repellent man, brought so vividly to life by the adept Kaplan that you almost feel you're in the room with him."   —Daniel Okrent, *Fortune*

"Having started sailboat racing myself in the summer of 1946, I understand the fervor that goes into it, so I opened Mr. Kaplan's book with a great deal of interest; I was not disappointed. . . . Tom Perkins might tell you that investing money in start-up companies is fun and—sometimes—profitable. But sailing has been what has truly obsessed him, and he seems to have succeeded in building the boat of his—and plenty of other sailors'—dreams."

—Pete du Pont, former governor of Delaware, *Wall Street Journal*

Damien Donck / *Newsweek*

About the Author

DAVID A. KAPLAN is a senior editor at *Newsweek* and the author of two prior books. *The Silicon Boys*, about the history and culture of Silicon Valley, was a national bestseller and has been translated into six languages. *The Accidental President* is the much-cited account of the controversial 2000 presidential election. Kaplan's cover stories for *Newsweek* have shown a similar range of interests, from baseball and Hollywood to capital punishment and American politics. His cover story in September 2006 broke the Hewlett-Packard corporate spying scandal that led to Congressional hearings. Please visit his website at www.davidakaplan.com for more information.

MINE'S
BIGGER

ALSO BY DAVID A. KAPLAN

The Silicon Boys

The Accidental President

MINE'S BIGGER

THE EXTRAORDINARY TALE OF THE WORLD'S GREATEST SAILBOAT AND THE SILICON VALLEY TYCOON WHO BUILT IT

David A. Kaplan

HARPER

NEW YORK • LONDON • TORONTO • SYDNEY

HARPER

INSERT PHOTOGRAPHY CREDITS: pages x–xiii, courtesy of Perini Navi; 1, 9 (top and bottom), 10–11, 14 (top), 15 (top), Giuliano Sargentini (courtesy of Perini Navi); 2 (top), reproduced with the permission of the Cutty Sark Trust; 2 (bottom), 3 (bottom), Tim Wright—photoaction.com; 3 (top), 5 (top left), Carlo Borlenghi (courtesy of Perini Navi); 4 (top), Franco Pace (courtesy of Royal Huisman Shipyard); 4 (bottom), Julian Hickman—1blueharbour.com; 5 (top right and bottom right), Justin Ratcliffe—ratcliffe.justin@gmail.com; 5 (bottom left), Emilio Bianchi (courtesy of Perini Navi); 6 (top and bottom), Dijkstra & Partners Naval Architects; 7 (top), 8 (bottom), Jed White; 7 (bottom), 8 (top), 16 (bottom), George Gill; 12 (top and bottom), 13, 14 (bottom), 15 (bottom), 16 (top), Bugsy Gedlek.

A hardcover edition of this book was published in 2007 by HC, an imprint of HarperCollins Publishers.

HarperCollins books may be purchased for educational, business, or sales promotional use. For information please write: Special Markets Department, HarperCollins Publishers, 10 East 53rd Street, New York, NY 10022.

FIRST HARPER PAPERBACK PUBLISHED 2008.

Designed by Kris Tobiassen

The Library of Congress has catalogued the hardcover as follows:

Kaplan, David A.
    Mine's bigger : Tom Perkins and the making of the greatest sailing machine ever built / David A. Kaplan.—1st ed.
        267, [16] p. of plates : col. ill. ; 24 cm.
    Includes index.
    ISBN: 978-0-06-122794-3
    ISBN-10: 0-06-122794-3
        1. Perkins, Thomas J.   2. Maltese Falcon (Yacht).   3. Yachts.   4. Yachting.   5. Clipper ships—21st century.   6.   Sailors—United States—Biography.   7. Businessmen—United States—Biography. I. Title.
VM331 .K213 2007
623.822'3B 22   2007280552

ISBN 978-0-06-137402-9 (pbk.)

08 09 10 11 12 ID/RRD 10 9 8 7 6 5 4 3 2 1

For Audrey

"The true peace of God begins at any spot
1,000 miles from the nearest land."

—JOSEPH CONRAD

LONGITUDINAL VIEW

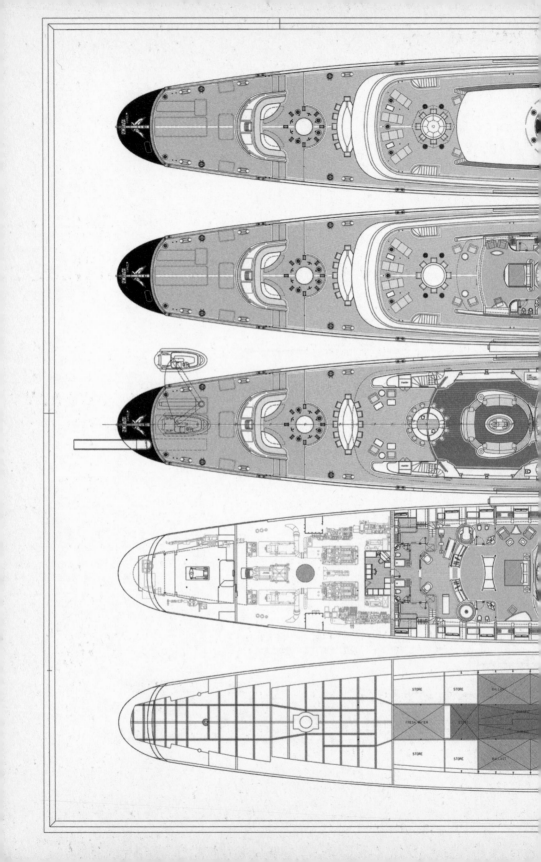

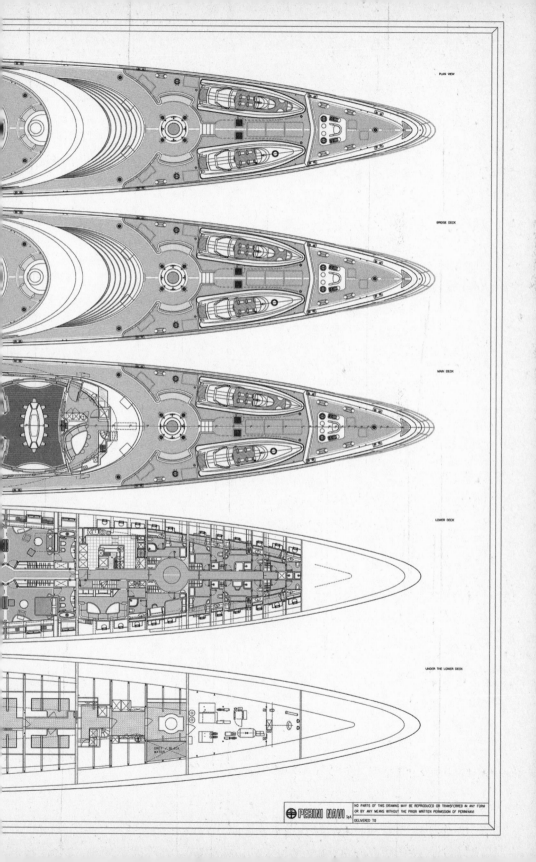

PLAN VIEW

BRIDGE DECK

MAIN DECK

LOWER DECK

UNDER THE LOWER DECK

GREY / BLACK
WATER

PERINI NAVI SpA

Contents

MINE'S
BIGGER

Making Waves

The sultans would have loved Thomas J. Perkins. And he would've felt right at home with them, at least until he decided to kick some sand in their faces when they didn't appreciate his sense of humor.

It was early in the summer of 2006, in Istanbul. On the left bank of the Bosphorus, before the marbled opulence of the Çirağan Imperial Palace, the government of Turkey was sponsoring a feast. Çirağan was the last palace built in Constantinople by the Ottoman Empire for royal family members. It was an historic location, dating centuries back—the waterfront home of High Admiral Kilic Ali Pasha in the late 1500s—with a commanding view of the strait that separated Europe from Asia, and commanding awe from any vessel passing by. In recent years, Çirağan had become a five-star hotel that the government used to entertain heads of state and other dignitaries. On this resplendent evening, all the culinary trappings to sate a sultan's appetite were present in abundance, though it's a good bet that Suleiman the Magnificent never saw such an ornate fountain of molten chocolate. The wrapped lamb dolma, the charcoal-grilled sea bass, the chicken galantine, the assorted sushi, the baklava—washed down with choice Doluca reds—were to honor an American and his spectacular new sailboat built entirely in a Turkish shipyard just east of the city.

The guest was Perkins, a venture capitalist who had bankrolled the modern Silicon Valley and made it part of the American imagination. If you had an entrepreneurial dream, or if you wanted to strike it rich—or better yet, if you wanted both—you headed for northern California. Over the course of thirty-four years, his eponymous firm Kleiner Perkins Caufield & Byers had become the Medici of the Valley—the most celebrated money machine in American history that most ordinary folks had never heard of. Perkins's firm funded such nascent companies as Genentech, which gave birth to the biotech industry; Netscape, which launched the dot-com boom; and Google, the darling of the Internet age.

In the risky business of providing start-up capital to fledgling entrepreneurs—and backing enough big winners to offset all the losers—Perkins performed the alchemy of turning millions into billions. Almost single-handedly, he transformed the art of venture capital—from the passive hobby of dilettante bluebloods into a cutthroat, hands-on profession that produced a generation of Siliconillionaires. Perkins became the man to see in the Valley. In the process, he had become fabulously wealthy himself and amassed great power. Along the way in his larger-than-life life, he'd managed to be a father figure to Apple's Steve Jobs, sailing mentor to media mogul Rupert Murdoch, and the occasional muse to romance novelist Danielle Steel—not to mention being flashed once by Naomi Campbell on a fashion runway and dining with Sophia Loren at his Bay Area home. He even managed to get himself convicted of manslaughter in France.

Now, at seventy-four, Perkins was setting out to transform the art of sailing. His $130-million yacht, anchored a few hundred yards out in front of the palace, was the *Maltese Falcon,* a twenty-first century clipper ship that was bigger, faster, higher-tech, more expensive and riskier than any private sailing craft in the world. The *Falcon* was as long as a football field, forty-two feet wide, twenty feet deep, with three masts, each soaring nearly twenty stories toward the heavens. On each mast were six horizontal yards—ranging from forty feet to seventy-four feet in width—to support the sails. The size of the *Falcon* was utterly out of scale with anything nearby—the ramshackle fishing boats, the tourist ferries traversing the Bosphorus, even the palace. If the ship were anchored in New York Harbor, she'd reach up to the level of the tablet carried in the arm of the Statute of Liberty. The exterior had teak decks, a varnished cap rail, stainless-steel fixtures and exquisitely finished surfaces—attributes of a classic ship—yet the overall look was sleek, metallic and ultramodern to the point of seeming foreboding. If Darth Vader had an intergalactic yacht built for himself, this is what it would look like.

On this day before the *Falcon's* 1,600-nautical-mile maiden voyage out of Istanbul and westward across the Mediterranean, the Turkish government was paying due tribute to Perkins, whose boat project was a milestone. It had kept hundreds of workers employed for more than a million man-hours over five years, and it had given Turkey's luxury-boat shipyards an international visibility they long craved. The reception was supposed to be hosted by the Turkish prime minister himself, but he was attending the funeral of his brother. Other government ministers, as well as naval commanders, came en masse. Their speeches lauded Perkins and beheld the 289-foot-long *Falcon* before them. The TV news cameras rolled and the local journalists seemed to write down every word. Even the leader of the political opposition, socialist Deniz Baykal, was a capitalist tonight, for the *Maltese Falcon* was the toast of the city and the country. "A lot of people asked me the reason I'm attending this gala," he said, wine glass in his hand. "Turkey has a great dynamism. In all sectors, big investments are going on. Yacht construction is one of these sectors. This yacht is a pioneer."

Now it was Perkins's turn to talk. He already had some sport with the event, unable to resist the chance to tweak both himself and Baykal in particular. The jet-black *Falcon* was not only all lit up by halogen lights, her fixtures freshly polished and her sixteen-member crew in dress uniform. Perkins had decked out the square-rigger in full plumage with dozens of signal flags—running from bow to stern, across the tops of the three giant carbon-fiber masts. Any guest would assume the *Falcon* was merely festooned with random, colorful pennants. But in the system of international maritime signals, each flag represented a letter. Ever the paragon of capitalism, Perkins's playful message spelled out: "Rarely does one have the privilege to witness vulgar ostentation displayed on such a scale." Perkins loved his *Falcon*—"the Big Bird," he had taken to calling it, though to some it looked more like a strange duck—yet he was surely aware that some of the locals may have deemed it excessive. To them, he wanted to get in a shot, even if they wouldn't get it since they didn't know nautical-speak.

And then he had a bigger dig, one that everybody would understand. In his brief remarks, delivered with a few well-rehearsed Turkish sentences spliced in, Perkins praised the work ethic of the Turks. "This yacht will stand up against the craftsmanship of any of the great shipyards of Europe," he said, wearing a black shirt and his trademark white linen suit from Macy's. "Some questioned the wisdom of my decision to build the boat here. They said the *Falcon* was too big, too complicated, too much for the Turks. I was told it couldn't be done to my standards and time-table." He proclaimed that had the *Maltese Falcon* been built in Italy—like his two prior superyachts—he wouldn't have been able to set sail for years, given the Italians' insistence on a thirty-five-hour workweek and the month of August off. But then Perkins added that he'd been annoyed for weeks that a shipment of fine sheets for the *Falcon*'s staterooms and china for the dining table had been held up for weeks by Turkish customs. Among other things, officials apparently wanted proof the British porcelain didn't pose a threat to human health, which raised the question how the identical merchandise was selling in retail outlets all around Istanbul. "This is what gives Turkey a 'Third World' reputation," he said in a tone that sounded both supercilious and oddly helpful. "There is no way this should be happening." Before Perkins finished, Baykal and others in the embarrassed government entourage were on their cell phones, calling the airport to see if the customs shipment could be located. Sure enough, at two the next morning, the sheets and china arrived at the palace and were taken by high-speed tender to the waiting crew aboard the *Falcon*. "He is an extraordinary gentleman, isn't he?" observed a Turkish businessman while witnessing the frantic phone calls that Perkins had instigated. "But I guess you shouldn't screw with him."

So it seemed.

After the shipment arrived from customs, we lifted anchor and docked alongside a tanker to take on twenty thousand more gallons of diesel fuel. We then set out for the ten-day shakedown cruise through the Dardanelles, into the Aegean Sea and the Mediterranean, on to Malta and up to the Côte d'Azur and the Italian Riviera. At that point, the *Maltese Falcon* had sailed only one day in her life. This was unusual: sea trials for a new yacht typically last for weeks to ensure that everything works and that the boat won't sink under the stress of wind and sea. But the two eighteen-hundred-horsepower engines performed flawlessly, the electronics showed only minor hiccups, the experimental sailing rig was still standing, and nothing leaked. And Perkins was in a hurry. He wanted out of Turkey, where he had spent parts of five years while overseeing the *Falcon*'s construction—often living aboard a motor yacht he kept docked at the shipyard. He'd had enough of the noise of all-night welders working on freighters the next dock over, toxic fumes from the painting shed, metal scraps in his scalp—and the nasty stray shipyard cat he named Satan. Perkins wanted the open sea.

The sea was his sanctuary—all the more so after the death of his beloved wife of thirty-three years, Gerd Thune-Ellefsen Perkins, a decade earlier. Perkins knew the Mediterranean and Caribbean better than his own San Francisco Bay. Untethered to the worries of everyday existence, life on the water seemed to allow Perkins to escape to a world of beauty, freedom, and on-demand solitude. He could brave the elements, yet live in the rarefied luxury of mahogany and cabernet. He could test out every new technology and gadgetry that an engineering geek loved, yet have chefs and stewards to cater to every detail of his needs. His old schooner, the 138-foot *Mariette*, had given him the chance to compete on the European racing circuit with a classic yacht. His 154-foot *Andromeda la Dea* allowed him to cruise the oceans and circumnavigate the globe. On that ketch, he had sailed to Antarctica before rounding Cape Horn, criss-crossed the Atlantic seven times, and in a victory of happenstance over prudence, survived the "perfect storm" of 1991 near Newfoundland that killed at least twelve people.

But even with the launch of his glorious new sailing creation, Perkins could not escape the other universe he had inhabited for the prior thirty-four years—the one that had made him a prince of Silicon Valley and a millionaire roughly five hundred times over. (The press routinely referred to him as a billionaire, but it wasn't so. He liked to say that accounts of his wealth became more and more exaggerated as the celebrity of the *Falcon* increased.) He wasn't only a financier and talent scout for start-up com-panies—nor just a confidant to dozens of CEOs and market-movers on Wall Street and beyond. Tom Perkins was a behind-the-scenes powerbro-ker, negotiator, and at times a deus ex machina in countless deals. As the *Falcon* unfurled her fifteen enormous sails as we passed by the great Hagia Sophia mosque of Istanbul, the city ablaze in tulips—while Perkins basked in the honking horns of onlooking boats and the handshakes of his guests onboard—he was cogitating over an escalating corporate board-room battle back home in northern California that he himself had trig-gered. It was a confrontation in which, by summer's end, he was no longer a background player. Instead, his name would be on the front page of every major newspaper in America.

The company was Hewlett-Packard, once the bastion of ethics in Sili-con Valley. HP was about to become immersed in a spying scandal that would lead to its chairman of the board, and other officials, getting indicted by the State of California on felony charges. There would be a civil settlement of $14.5 million and Congressional hearings as well. Per-kins had been on the board of directors, but quit in a fury just a month earlier. The public had no idea why he left. An HP press release merely stated he had resigned and that the company, in boilerplate language, thanked him for his years of "service and dedication." In fact, he had quit because the HP chairman, Patricia Dunn, had authorized spying on board members. She supervised efforts by private investigators to obtain directors' personal phone records and even to trail some of them. Her goal was to root out who was leaking information to the press from the boardroom—a chronic problem at HP dating back a few years to when the controversial Carly Fiorina was running the show. Because of corporate confidentiality requirements, Perkins believed he could not immediately go public himself with the reasons he left, but would spend nine weeks trying to force HP itself to disclose the reasons.

Depending on who was narrating the tale, Perkins would be the white knight, or dark lord, of the story—the man who blew the whistle on corporate malfeasance or who turned a personal score into a high-profile vendetta. But all sides agree that it was he who disclosed HP's sub rosa surveillance of its own directors, as well as journalists—and that this public spotlight spawned the scandal, the prosecutions, and the humiliation of a legendary company. Dunn, the besieged HP chairman, learned first-hand what the Turkish businessman at the Imperial Palace surmised about mixing it up with Perkins.

Yet Perkins hardly reveled in it. He was ambivalent. It was one thing to have his way with Turkish customs, quite another to orchestrate the takedown of a woman whom he regarded as a pathetic lightweight. It hardly helped that she was suffering from advanced cancer, an illness that had claimed his own wife. He viewed himself as a righteous intervener, and he was thrilled when others saw it that way, yet winced at criticism that he was a steamroller. Perkins had an ego and the conviction that people in his league usually do—that he was invariably right. But he wasn't merciless and he wasn't a jerk. And he knew that his two worlds of high risk—sailing and finance—were colliding. He wasn't sure he wanted it to be so.

Day by night, as he experimented with the *Maltese Falcon*'s sail formations and angles of attack on the wind, Perkins absorbed himself in his technological masterwork. Most of the time, he seemed to prefer that obsession to the corporate intrigue he thought he'd long rid himself of. No longer was he on the board of thirteen public companies at the same time—chairing a record three of them on the New York Stock Exchange. The *Falcon* was the crowning achievement of his half-century of yachting. It was to be his retirement project—actually, given the boat's mammoth size, it was to be his self-described "retirement *village*." As the maiden voyage got under way that morning in July 2006, Perkins decided he'd try his damnedest to stay on course with what was supposed to be the trip of his life.

It was a journey long in the making. Perkins had gotten what he wanted most of his life. The *Falcon* was the ultimate prize—a boat nobody else could have conjured up, a fantasy that made him a visionary, a fool, or both. Nearly six years earlier, he had decided to create what he called "the perfect yacht." Given both advances in materials science and the explosion of dot-com lucre, two other American tycoons were attempting about the same thing. All the better to Perkins: he wanted to make the best boat, but it would be even more satisfying if he could beat out others while accomplishing it. "Mine's *bigger*," as Perkins liked to say. Of course, that meant somebody else's needed to be *smaller*. In a kingdom of haves, he had to be a have-more. Temperamentally, constitutionally, pretty much clinically, Perkins needed to be in a battle of egos on the high seas. After all, there's not much fun in winning a competition of one.

Perkins's "clipper yacht" was intended to evoke the era of magnificent vessels that once raced across the oceans. But his 1,367-ton square-rigger would be more New Old Thing than mere tribute to the past. It was a futuristic marvel born of modern technology and design. Gone was all the rigging: there were no ropes, no wires, nothing to support the masts or the horizontal yards, or to control the fifteen sails carrying nearly 26,000 square feet. No longer was there a score of deckhands to climb the rigging to furl and unfurl every sail. Instead, the masts were entirely freestanding and, unlike masts on any other boat, they were not stationary, but rotated. The sails were deployed at the push of a button, rolling out from inside each twenty-five-ton mast. Dozens of computers and microprocessors—connected by 131,000 feet of cable and wires—integrated the system, allowing helmsman and crew to control the boat nearly effortlessly. And unlike the clippers of yore, with their vast, white expanses of billowing canvas, the *Falcon*'s sails in effect formed a nearly flat vertical wing on each mast. Conventional clipper morphs into fin-de-siècle machine—a marriage of old and new, or a mutant of tradition? It depended on your perspective.

Damn the risks, damn the uncertainties, damn the costs—it was full speed ahead. This was the Silicon Valley ethos that Perkins was so elemental in establishing. Now he would change the culture of sailing. Part art, part science, and part magic, sailing was a way of commerce for thousands of years. It had been the instrument of global discovery going back to antiquity. It was a romanticized sport of kings since the seventeenth century. Sailing was no longer necessary to travel the world, but it remained essential to Perkins. Sailing was beautiful, dangerous, enduring, primordial, noble. Sailors reined in nature and harnessed the wind—yet were at their mercy. Tom Perkins resolved to leave his mark on that long arc of history and imagination—in short, he intended nothing less than a sailing revolution. And a vessel through which his boundless ego could be expressed—the largest privately owned sailboat on the planet, the *Maltese Falcon*.

This is the story of that yacht and the man who built it.

ONE

Killer Instincts

In the warm sea breeze of early autumn, amid the splendor of the south of France, who knew death was also in the air?

It was the first October weekend of 1995. In the small turquoise bay of Saint-Tropez, Tom Perkins was king of the sailing world. The American venture capitalist was racing his two-masted schooner, *Mariette*, in the La Nioulargue, an annual regatta that drew thousands of spectators and hundreds of boats to this once-medieval sailing village of the Riviera. Though perhaps surpassed by other jet-setting playgrounds on the Mediterranean, Saint-Tropez was still mythic for bronzed movie stars, fast Ferraris and a Dionysian beach culture. For his part, Perkins—at sixty-three, no young playboy—was hardly the embodiment of glitz. But the Nioulargue was glitz—with its spectacular array of multimillion-dollar period yachts and the gold-diggers and gawkers who wanted to get close to them. The week of races capped the European classic-yacht racing circuit and was the last dance of the Mediterranean summer. While the regatta didn't carry the status of the America's Cup or the Newport-to-Bermuda competition, the Nioulargue was about bragging rights and showing off.

The steel-hulled *Mariette* was brilliant. Including the bowsprit, she was 138 feet long. Built in 1915, *Mariette* was deemed one of the masterpieces of Nathanael Greene Herreshoff, the legendary naval architect. The

MIT-trained Herreshoff—the "wizard" of Bristol, Rhode Island—designed yachts that were recognized for their grace in the water and below-deck elegance. His clients were the barons of his day: William Randolph Hearst, J. P. Morgan, Jay Gould, Cornelius Vanderbilt III, Harry Payne Whitney. Yet the dark-blue *Mariette*, built for New England wool merchant Frederick J. Brown, was seen even then as a yacht above all others, boasting a stunning Edwardian width-of-the-boat walnut-paneled saloon that was the envy of any captain of industry. (The bathtub in the owner's cabin, then a rare luxury on yachts, was notable, too.) *Mariette*'s gaff rig was renowned along the north and south shores of Boston. Even today, with her remarkable complement of thirty-eight cream-colored sails, and hand-carved scrollwork on the bow and stern (done by Perkins himself), she remains one of the most-photographed boats in the world. On glossy calendars, she is to yachting what Raquel Welch used to be to bathing suits. Perkins had rescued a deteriorated *Mariette* in 1995 for about $6 million. She was one of three big boats Perkins owned in the 1990s, but she was the project to take his mind off the recent death of his wife Gerd.

Befitting a French sporting event, the Nioulargue was known among sailors as an anarchic event. Nobody confused the buttoned-down, ship-shape orderliness of the New York Yacht Club with the free spirits running a regatta right off the shoals of Saint-Tropez. *Vive la liberté!* Courses were poorly designed, race starts were frightening, and there were just too many boats occupying too little space on the bay. The last miscalculation came with the territory: since many spectators watched from shore, rather than aboard party vessels or helicopters, the starting line had to be near land—indeed, one end of the line was the rock bearing the name Nioulargue. But the other two attributes of the race were simply classic French disorganization. The race consisted of two divisions—one for the smaller boats (roughly thirty feet and over) that numbered in the hundreds, and one for the twenty-five larger yachts like *Mariette*. The course was a triangle, with boats required to go around the triangle twice. The smaller—and slower—boats began their part of the race first. Trouble was, as those boats were finishing their first lap, they were passing through the starting area that was being used by the larger yachts to begin their part of the race at that moment. It was a traffic pattern designed for disaster. In the past, it had meant only minor collisions, near collisions, and the occasional cry of *"Sucez la pipe!"* near the starting line. This year would be different.

Going into the last day, *Mariette* led the regatta in points. In the final race, at the line, Perkins found himself maneuvering for position with the other big boats—as well as scores of vessels, both power and sail, in the spectator fleet that were jockeying to see the start. It was nautical pandemonium. Several helicopters, with photographers for the trade magazines, roared above. The breeze was fresh and *Mariette*'s crew of thirty-two (not including Perkins's personal chef and six guests) had all it could handle. Despite having a professional captain on board (and Herreshoff's grandson Halsey), Perkins was himself at the helm—as he usually was, especially at the ever-critical starting line. Alone among owners of the million-dollar yachts, he drove his own boat. Intense, impatient, stubborn—Perkins could be a hell-raiser on the water, a contradiction to the lineage of gentleman sailors that had produced classics like *Mariette*. When he wanted lunch "in twenty minutes," the crew knew it meant five. When he asked for navigational data "as soon as possible," it meant yesterday. Perkins fulminated, he gesticulated, he sighed with deliberate drama, he famously stomped on his Dunhill hat. Often, he did them all in reaction to crisis, real or otherwise. The hat-dance led some of his friends over the years to refer to him privately as Rumpelstiltskin.

In a sport that was subject to the vicissitudes of nature—wind, currents, storms—Perkins was a control freak. He wanted people, technology, things, to do what he asked, to be what he expected. He had the linear mind of an engineer through and through. He pretty much had the same breakfast of eggs and fruit every morning. When he had a chocolate croissant in 1996, the crew on *Andromeda* was so shocked it noted the occurrence in the ship's log. When he couldn't find the exact kind of harpsichord he wanted for his home, he constructed one himself. When he often woke up in the middle of the night with insomnia, rather than

dipping into a magazine, he'd read a biography, like that of Austrian logi-
cian Kurt Gödel who devised the Incompleteness Theorem. Perkins loved
that kind of book not just because it was intellectually challenging, but
because it ratified his view of things—that there was a transcendent,
objective mathematical reality that underlay everything and that it existed
regardless of human thought. That kind of certainty was appealing in the
face of pesky fortuities. Even when acts of God took hold, Perkins looked
to analyze his environment, which gave him the semblance of control.
When he found himself in the middle of the shaking, twisting, undulat-
ing Golden Gate Bridge during the 1989 Loma Prieta earthquake, he
thought less of his own mortality than "How many 'degrees of freedom'
does this bridge have?" (He got to twelve before concluding he ought to
start worrying about his own mortality.) "This is sick," he thought to
himself. "You are such a goddamn engineer."

Yet machines could be trusted. For a control freak, people were the
hardest to control. In the land of venture capital, he often said that it was
individuals who caused the problems and only rarely the technology. On
the water, nothing bothered Perkins more than guests who got seasick.
He thought *mal de mer* was a character flaw—it seemed apt to him that
the French had a quaint term for it—a sign of weakness rather than just a
bum middle ear. Some years earlier, on a different yacht that Perkins
owned, a friend became queasy. But she had heard Perkins trashing someone else who had gotten sick on the rolling sea. Terrified, she excused
herself and vomited in her Gucci purse before she could make it to the
head. One of the small secrets of his life was that once—just once, in the
English Channel—he got seasick himself.

Perfection was something he particularly expected during a regatta.
If Perkins lost a race, heaven forbid, pity the person who incurred his
frustration, even if the person had nothing to do with the defeat. Once,
the designer Count Hubert de Givenchy was visiting Perkins on *Mariette*.
At the time, Perkins was dating Danielle Steel, the bestselling American
romance novelist, who was aboard with him. Though Steel and Givenchy
were close friends, the count had not met Perkins before. Perkins had lost

a race that afternoon, but he knew that Givenchy was still due the honor of being personally picked up on shore to go out to the anchored *Mariette* rather than having a crew member drive the tender. Too bad for Givenchy, then pushing seventy and recovering from pneumonia. As Steel recounted the story, Perkins was still so mad about his defeat that he bounced Givenchy in rough seas all the way from the dock to *Mariette*, "drenching Hubert totally." An arbiter of fashion and manners, Givenchy was a good sport about it, though he had to remove his clothes and have them dried when he got to the boat. "With a twinkle in his eye," Steel said, "Hubert told me, 'I don't usually take off my underwear when I come to dinner.' It was the first time he met Tom. As Hubert left the boat, he hugged me and said, 'My dear, you are going to have your hands full.'" When Perkins and Steel went to their cabin that night, she lit into him. "I don't believe in spankings," she scolded him, "but if this is the way you behaved as a child, with temper tantrums like that, then you could've benefited greatly from a good one." Perkins was stunned by the reprimand, and apologized to Givenchy later on.

For a person who had made his fortune through the exercise of cold-blooded, clinical judgment, Perkins's occasionally ferocious, intemperate displays seemed out of character. He acknowledged it, sort of, for he could poke fun at himself for these outbursts, sometimes minutes afterward. He often said that the only thing bigger than his yachts was his ego. And amid all of *Mariette*'s mischievous practical jokers, he could always top the crew's plans with elaborate schemes aimed at other yachts. Most of his professional crew thought he was fair and approachable and honorable—traits that didn't fit the typical owner of a superyacht.

Perkins didn't look the part either. He was tall and tan but not overly vain, making only a modest effort to cover his graying hair, even though he was then into his sixties. Other owners might attempt the Thurston Howell III air: ascots and blazers, pursed expressions, aloof, never to be called by a first name. Perkins wore white New Balance sneakers and metal-frame glasses that had gone out of fashion during the Eisenhower administration. He walked about the boat with long strides, in a hurry to get to what was next. He ate with the crew when no guests were on board. He laughed as frequently as he sighed. And Perkins loved to mess with pretenders to his circle. When one of his yachts was at a dock and a wannabe came by insisting he knew the owner, Perkins would introduce himself as a mechanic and inform the stranger that the owner just couldn't be disturbed right now. When King Juan Carlos I of Spain came by *Andromeda* once, Gerd had to call Perkins down from the mainmast to meet him, a greasy wrench still in her husband's hand.

If you took Perkins away from the wheel of his glorious ship, you might think he was just another MIT geek. The crew may have feared him at times—he did sign their checks and he was responsible for giving them the job of their young lives—but they liked working for him. Every one of them called him Tom. How to square this jumble of attributes? Perkins just wanted things his way. He had the mindset of an engineer, yet understood human imperfections. He read technical manuals, but soaked up literature and history. In short, for a gearhead worth many millions, he was usually a kick to be around.

Except in a regatta. Under normal conditions, Tom Perkins behaved at the start of a race like a shark that hadn't been fed in a week. The start of a sailboat race frequently determined the overall victor, and jockeying for the best position made it the hairiest time of the contest. But in La Nioulargue, circumstances were worse. All the yachts were moving at ten to fifteen knots. That may not sound fast, but this was open blue water, not a highway with defined lanes. The scene was orderly chaos, akin to a latter-day Dunkirk. *Mariette* was in a particularly precarious position. She was sailing on port tack—meaning that the wind was blowing across the port (left) side of the boat and her sails were on the starboard (right) side. Under well-settled maritime rules of the road—in a race or otherwise—a sailboat on starboard tack always had the right-of-way over a sailboat on port tack. So *Mariette* had to steer clear of any sailboats coming at her from starboard. Another boat on port tack was the thirty-six-foot sloop

Taos Brett IV, approaching from a sharp angle. Perkins assumed she was part of the spectator fleet rather than what she was—the lead vessel from the first division that had started more than an hour earlier. He also assumed the smaller boat would get out of the way and, being a hyper-aggressive helmsman, he wasn't about to give in. But *Taos Brett IV* wasn't altering course and instead was plowing her way by all the big yachts crowding the starting line for their own race. Her skipper wanted to win no less than Perkins did.

With both boats on the same tack, the rules dictated that the leeward boat—the one furthest from the wind (*Taos Brett IV*)—prevailed, *unless* the windward boat (*Mariette*) was obstructed (for example, by other boats or by land). In that event, the leeward boat had to give way. It might sound like these universally accepted rules and exceptions were drafted by a committee of seersuckered lawyers who couldn't possibly appreciate that skippers had to make split-second judgments. But the rules actually were based on some semblance of common sense, as nobody wanted to see multi-ton, fast-moving craft slam into each other. Still, the traffic rules were subject to interpretation, and in close situations, two skippers conceivably might each be a little right. Which unfortunately meant that they each might also be a little wrong. In this close situation, *Taos Brett IV* was the leeward boat and her helmsman believed he had right-of-way. But *Mariette* had two huge yachts bearing down on her from both sides—coming from the opposite direction, on starboard tack. Perkins could not turn in either direction without ramming into one of them. With *Taos Brett IV* nearly upon him, Perkins's only hope was for her to realize that *Mariette* was obstructed.

Not only did *Taos Brett IV* arguably lack right-of-way, she was about to violate the oldest axiom in the sailor's handbook—better to veer off than to end up on the bottom of the sea gurgling, "But I had the right of way!" That's especially so when your boat is less than a quarter of the size of the other vessel coming at you. Perkins's panicked crew, as well as crew on other boats nearby, began bellowing at the French skipper, Claude Graff, to get out of the way. Standing on the high side of the heeled-over

Taos Brett IV, Graff saw exactly what was happening. But according to Perkins, Graff gave him the finger and continued to hold his ground, trying to slip by *Mariette's* bow with just enough room to spare. Perkins kicked his two diesel engines into reverse (as a precaution, big yachts always keep their engines on, in neutral, when starting a race), and reduced *Mariette's* speed to a few knots.

Taos Brett IV almost cleared *Mariette's* bow, avoiding a direct hit. But the schooner's long bowsprit clipped the rigging on the port side of *Taos Brett IV*, and began carrying the smaller boat along—and then under *Mariette's* bow. The rigging was one of the wire stays that supported *Taos Brett IV*'s mast and the wire failed to snap. Within two seconds, *Taos Brett IV* capsized, filled with water and sank. "We didn't damage the boat—we just sank it," Perkins recalled.

The doomed sailboat had five sailors aboard, three in the cockpit and two in the cabin who were packing a spinnaker for use in the race. The pair inside had no idea what was happening in the duel between *Taos Brett IV* and *Mariette*. One of those people—a fifty-year-old physician from Nice named Jacques Bourry—didn't make it out alive. Horrified at the collision and Bourry's drowning, Perkins immediately dropped out of the race, which was soon canceled anyway—as was the regatta itself, for several years, because of the tragedy. Perkins was so shaken he gave the helm over to his captain, Tom Eaton.

Forty-five minutes later, after *Mariette's* sails had been furled and stowed, Perkins was arrested by armed French authorities and charged with manslaughter. He didn't learn until midnight—ten hours after the collision—that *Taos Brett IV* was actually racing and in the lead in her division, and that the race committee had idiotically set up a course that brought the smaller boats directly into the starting line of the bigger ones. Eaton was arrested as well, even though he wasn't at the wheel. So, too, were the skipper of the other boat and the chairman of the race committee. Perkins was certain he had done nothing wrong—that the other boat had erred and that it was all a terrible accident—but he was nonetheless unsurprised at being taken into custody. "I was an American who had just killed a Frenchman—what did you expect?"

Evidently, Bourry's family agreed that it was an unfortunate accident. The next morning, his wife—as well as his mistress—came by *Mariette's* slip to see Perkins. They had just driven in to Saint-Tropez from Nice. He assumed they had come to berate him. Instead, over cappuccino and cookies, they hugged him and told him that he was not to blame for Bourry's death and, oh, one other thing—the widow wanted to know if Perkins would take on her teenaged son as crew the following summer. (Perkins felt he couldn't say no, though his lawyer subsequently told him that if he really hired the boy, Perkins would be looking for a new lawyer.) Notwithstanding the widow's forgiveness, the local French prosecutor wanted Perkins's hide. Bourry was much loved in Nice. Egging on the prosecutor was the French press, which was having a field day with photos of the collision. "MONSTER AMERICAN YACHT KILLS FRENCH DOCTOR!" is how Perkins remembered the end-of-the-world front-page headline in one of the Paris tabloids. Hyperbole aside, French law typically required a criminal inquiry in homicides, even if accidental. The sensationalist overlay of the regatta and its millionaires virtually ensured there would be a prosecution.

Perkins attended Bourry's funeral and, after he got his passport back from French police, he left the country. In theory, Perkins could've avoided prosecution by choosing not to come back. But he did return for the trial—exactly a year after the accident, timed to coincide with the anniversary—to be held in the old military village of Draguignan. While this was not the O. J. trial of France, the prosecutor was still scoring media points and kept saying he wanted Perkins to spend some time in the pokey: the American was big game and a risk-free target in the provincial countryside. Though friends gave him the names of top-flight criminal-defense counsel in France, Perkins underestimated the storm into which he was heading. So he hired a less prominent Parisian lawyer who specialized only in such maritime accidents as major oil spills (including the

Amoco Cadiz disaster off Brittany in 1978). It would prove to be a strategic blunder, as well as slightly embarrassing—for unrelated reasons—to his celebrity girlfriend at the time who went on to be his wife and then ex-wife.

———————

She happened to be Steel, who was French by background and upbringing, and was fluent in the language. Fifteen years younger than Perkins, she served as his translator at the trial, though few in the courtroom had any idea who she was, or of her publishing and fashion éclat in France and around the world. Perkins and Steel had known of each other from San Francisco society for two decades, and they began going out in 1995, a year after Gerd had died of cancer. A few days before the trial, Steel flew in from the United States to Paris, where she connected to Nice. Perkins's lawyer had come up to Paris to join her for the last leg, the better to plan strategy.

But the lawyer, according to Steel, seemed to be more interested in flirting. Shortly after she checked into her suite at the Byblos in Saint-Tropez, the lawyer came by—"this little French guy with spindly hairy legs," as Steel liked to derisively refer to him. He had been in a business suit on the plane and in the limousine, but now looked ridiculous, as if part of a Monty Python skit: he was wearing only a shirt and tie, pink French shorts, brown droopy socks—one barely clinging to his knee and the other scrunched around his ankle—and clunky business shoes. Steel couldn't figure out if she was more annoyed or amused. "He had this arrogant air, as if he were Cary Grant or Alain Delon knocking on my door. And no wine!" Steel sent him on his way.

At the trial, the lawyer—in his fancy powdered wig—sat right with Steel and the defendant. Perkins had audacity and gall, but was more scared than he'd been in his life—on trial for his freedom in a foreign country, before a panel of judges who sounded entirely ignorant of, and uninterested in, sailing rules or terminology. The lawyer knew Perkins was terrified, but couldn't resist a few more coos at Steel. Indeed, whenever she stretched out her arms during the hours and hours of trial, she found the lawyer's fingers reaching to touch her neck. Harmless as it was, the scene just added to the cultural and legal absurdity that Perkins found himself in.

Perkins's defense in the courtroom did not go well. The three judges—with no jury—were disdainful of expert witnesses and photographic evidence of the accident. They ruled all the sophisticated charts and re-creations of the collision inadmissible. While there had been at least a thousand witnesses, few accounts made it into testimony. And the judges seemed uninterested in the context of the accident—that it was a sailboat race. The prosecutor successfully argued that because Perkins acknowledged *Mariette* was going faster than the nautical speed limit—which was five knots within a third of a mile of the French coast—it was prima facie evidence he was behaving criminally. The argument was absurd, something Joseph Heller might've cooked up: the starting line of the race was not only within a third of a mile of the coast, one end of the starting line *was* the coast.

The small-town judges also appeared to be rather chummy with the prosecutor, whereas Perkins's lawyer was the interloper from the big city. The court would not even allow Perkins to testify in his own defense. "By the end of the first day," he said, "I'd lost interest in my own manslaughter trial." There he sat, under guard—all six-foot-two of him, silent, staring, stewing, powerless. It scarcely helped his cause that his dismissal of the French as dumb peasants was so transparent. Or that his defense wasn't able to personalize him—to paint him as a human being in addition to a shrewd and driven MIT scientist, Harvard MBA and Silicon Valley magnate. Not that it necessarily mattered. In the end—given that a man died, given that French criminal law was fairly strict in this arena, and given that the defendant was an American with very deep linen pockets—Perkins probably didn't stand a chance from the moment that *Mariette* drove *Taos Brett IV* into the sea. As the prosecutor put it in his damning closing, everyone had behaved like *"des loups de mer"*—like "wolves of the sea."

Steel had her own travails at the trial. Spectators on occasion came up to Perkins during breaks to complain of his homicidal ways. Steel had to translate. At one point, a French racing expert didn't let up in his stream of insults. He had no idea that Steel was also Perkins's companion. When she'd had enough, she stopped passing along the English version to Perkins—who said nothing—and instead she just lit into the French guy. This went on for a few minutes. Finally, the racing expert asked Steel, "What kind of translator are you anyway?"

After a three-day trial, the court convicted Perkins of involuntary manslaughter—excoriating the ruthlessness of sailing competitions—and found the other sailors guilty as well. But the judges did not heed the prosecutor's cry for prison time for any of them. Perkins received a sixty-day suspended sentence and a ten thousand dollar fine, barely a rounding error in his bank accounts. (Perkins paid his own captain's fine, too.) Apart from having to notify the Securities and Exchange Commission of his conviction—after all, he did still serve on the boards of all those regulated public companies—the conviction had no consequence in the United States. In contrast to the French news media, which went nuts, the American press curiously missed the story altogether; these were the ancient days of the 1990s, before Internet databases made it easy to keep tabs on any corporate official. For her part, Steel thought the whole affair would make for a good novel—a yachtsman in the dock for homicide in a land where he couldn't speak the language, couldn't testify, and was defended by a leering lawyer—but she wasn't one to exploit the misfortune of a friend.

All in all, Perkins lucked out. Bourry's family filed a civil claim against Perkins, which largely worked in Perkins's favor as well. While it ended up costing around fifty thousand dollars, his insurance carrier, Lloyd's of London, told him his liability would've been far greater had the incident happened in the United States. France may have had a stricter criminal code, but the price of wrongdoing was a lot less. Perkins was also fortunate because, according to Perkins's civil lawyers, Bourry had underreported his income. So any lost wages that a court awarded his family were below what in theory they should've been. A friend of Perkins told him, "You're so fucking lucky: you managed to kill someone, but you did it in France. It barely cost you." So meager was the money to Bourry's widow that when all the proceedings came to a close, Perkins mailed a check to her to cover her son's college education. In matters of finance and sailing and people and life, Perkins could be reckless and bold. But he could also display unusual grace. In a situation in which he might have gloated and bid the French nation a foul farewell, he found the decency to do a kind act on behalf of a boy who had lost his father.

Next to the death of his wife Gerd, which had sent him into despair for a year, La Nioulargue was as unpleasant an episode as Perkins had ever experienced. Until the obloquy of Saint-Tropez, he had led a rather charmed life. But the experience hardly left him chastened. To the extent that the moral of the trial—regardless of how legally unsound it might've been by American jurisprudential standards—was that Perkins may have been sailing too hard at the starting line, Perkins didn't get the message. It didn't even occur to him there might be a message to get. Why would there be? He had been willing to bet everything during his entire life—in sailboat racing, in venture capital, in business, in relationships. Like so many before him raised on the East Coast, he had gone west in search of adventure and treasure. Before he was thirty-five, he had risked all his savings to turn a laser invention into a business and to become a millionaire. Before he was forty, he chucked that success and a safe management career at Hewlett-Packard to create Silicon Valley's first venture-capital powerhouse.

At the helm of a sailboat, he was recognized as a bellicose if brilliant tactician who could squeeze both an extra knot out of his vessel and the gonads of any opponent. So, maybe he'd been too aggressive at the line in Saint-Tropez? Maybe there was a reason he always had to be at the big wheel of *Mariette*, even when other owners were content to let their professional captains drive the boat? Never mind any of that. Perkins lived for drama. Self-doubt was anathema to him, antithetical to the personality—or pathology—of that which made him such a success. Yesterday was gone. No, after the disaster of Nioulargue, after a man had died, after his conviction for manslaughter, there would be no looking back. Fearless as ever, Perkins had something altogether else in mind. *Mariette* was splendid. But Perkins wanted "the perfect yacht."

TWO

The New Old Thing

The history of boats is about as old as the record of mankind. It was into that long illustrious history that Tom Perkins longed to fit. Over many centuries, long before there were engines, the progression of experimentation—in size, rig, materials, and hull design—reflected the mundane demands of exploration, commerce, and warfare. But in the more complex arena of ego and prestige, innovation also signified a need to show off. In the way that twentieth century societies used Sputnik or the Saturn V to command respect, countries and men of a different era did so with ships. When Columbus landed in the New World, the first thing the locals saw were three great multimasted vessels with white wings, the likes of which they had never imagined. They believed they were seeing gods of the heavens approaching from the endless sea. Perkins was born centuries after the Age of Exploration, yet intended the *Maltese Falcon* to be that kind of awesome sight.

In the beginning, there were rafts, unless you counted holding on to a tree limb floating down the river. In time, the ancient seafarers produced hollowed-out hulls and other watertight vessels—made of reeds, timber, and skins, propelled by poles, paddles, and oars. Then came the first sail, perhaps in the South Pacific or Taiwan. The earliest image of a vessel using the power of wind—on an Egyptian vase—dates to more than five thousand years ago. It was a narrow dugout boat with a single square sail. Riverboats on the Nile could only "run" with the wind—the breeze behind them—and it could be slow going as the prevailing wind was usually in the opposite direction of the current. Archaeological finds of models and pictures show the Egyptians experimented with different primitive rigs: broad low sails and tall narrow ones; sails supported by poles at top and bottom; and sails supported only by a horizontal yard at the top. For many cultures, sailing carried with it much symbolism: the mast was a link between heaven and sea.

Over the course of centuries and among various civilizations, sails took on different shapes—typically variations on a triangle—and sailors learned to point closer to the wind. That meant their boats didn't just sail with the wind from behind, but with the wind blowing from near the front of the boat. So if the wind was coming from the "twelve o'clock" position, the boat might be able to move in the direction of ten o'clock or two o'clock. This newfound flexibility, as well as larger sails and larger craft, enabled voyages into the oceans by 2000 B.C. Homer's *Odyssey* contains stirring accounts of sailing in a Greek galley: "Telemachus shouted out commands to his shipmates: 'All lay hands to tackle!' They sprang to orders, hoisting the pinewood mast, they stepped it firm in its blocks amidships, lashed it fast with stays and with braided rawhide halyards hauled the white sail high. Suddenly, the wind hit full and the canvas bellied out, and a dark blue wave, foaming up at the bow, sang out loud and strong as the ship made way, skimming the whitecaps."

As vessels became larger, they included accommodations and storage areas below deck that made extended passages possible. Ships with more than one mast can be found as far back as 200 A.D., in records of the Roman Empire. More and more square-rigged merchant ships were built that surpassed two hundred feet in length. These were far too long to use oars for propulsion, and sail reigned supreme. A square-rigger was defined by its square sails set on yards placed at right angles to a vessel's hull—"athwart," or across, the ship, just the opposite of fore and aft. However, the "square" sails weren't precisely square—they were usually narrower at the top than bottom, with curved vertical edges. In northern Europe, by the sixth century, the Celts and Saxons had developed their own sailboats. The Vikings followed one hundred years later, and by the middle of the medieval period they had perfected their narrow, classic "longships" that traversed the Atlantic and roamed the Mediterranean. Some Scandinavian ships also introduced a critical improvement—the keel, which added stability, windward lift, and a seating in which to place a mast. Over time, deeper keels allowed for more stability (thus the phrase "even-keeled") and more sail area, and therefore more speed. The Vikings became avid raiders, explorers, traders, and settlers. In England, toward the end of the Middle Ages, Henry V's square-rigged flagship *Grace Dieu* was the greatest of all ships when she was launched in 1420. She was 218 feet long, with her mainmast almost that tall.

Not all medieval countries used square rigs. Far from the North Sea, a different tradition arose in the Mediterranean—the lateen rig (Latin for "sail"), which utilized one or more triangular sails. Unlike on a square-rigger, whose yards and sails were positioned *perpendicular* to the length of the boat, a lateen sail ran in a fore-and-aft direction—from the bow to the stern, *parallel* to the keel. The lateen sail was mounted on a long wooden spar that was attached diagonally to a short mast. The spar's position was very different from that of a boom attached at a right angle to the mast, like on a conventional sloop or ketch or yawl, which wouldn't be invented for hundreds of years. (The modern Sunfish, the most popular recreational boat ever, is a lateen, but with *two* spars off the mast supporting the single triangular sail.)

Different sailing architectures were developing in Europe and Asia, due in part to happenstance, and in part to regional wind conditions that facilitated different commercial and military agendas. Different rigs had different advantages and disadvantages. A square-rigger could run with the wind, but couldn't sail close to it. Lateens did better sailing upwind and might require fewer deckhands to handle, but half the time on some lateens the sail was on a "bad tack": because the sail was set to the side of the mast, rather than behind it or beneath a yard, sometimes on one tack the sail pressed against the mast, which interfered with smooth airflow over the leeward side of the sail.

————

At the turn of the fifteenth century, sailing was the only means of going any distance by sea. The variety of rigs across Europe and Asia reflected a largely unconnected world. That changed with the ascendance of European powers that were eager to expand their economic and military reach, and had the shipbuilding expertise to do it. These were the early days of the taller three- and four-masted sailing ships that immortalized explorers like Columbus and Magellan. These new craft represented a technological sea change in ship design that would change civilization. While they were primarily square-riggers, they also often carried lateen steering-sails that helped the ship to maneuver close to the wind and to tack through it. ("Tacking" was changing course by going from port tack to starboard tack or vice versa). Additional masts allowed for more and smaller sails. That meant less strain on both crew and equipment, as well as permitting more combinations of sails to optimize wind and sea conditions.

The three Spanish ships of Christopher Columbus—the fabled *Niña*, the *Pinta*, and the *Santa Maria*—were all well under one hundred feet long, small even by the standards of 1492. On his epic voyage to the New World—the Spanish Main, as it was called—Columbus happened upon the Atlantic Ocean's circular wind pattern: northeast trade winds off the coast of Africa running by the Canaries and all the way to the Caribbean, but westerlies further up that reversed direction. Columbus followed the trade winds: that's why he wound up so far south—in the West Indies rather than Weehawken. Though there could be doldrums or storms, this wind pattern prevailed enough that the ships of the Age of Exploration learned they could sail with the wind behind them all the way to the Americas, and then again back to Europe. That made the square-rig design particularly advantageous.

Six years after Columbus's initial passage to the West, Portugal's Vasco da Gama headed in the opposite direction and became the first European to reach India by sea. Then, in 1519, Ferdinand Magellan—another Portuguese navigator but sailing under the Spanish flag—began the first successful circumnavigation of the globe. Over a period of three years, his men traveled across the Atlantic and around Cape Horn at the tip of South America, entered the Pacific en route to the Philippines and on to the Cape of Good Hope around southern Africa and back to Spain. While Magellan was killed along the way, the flagship made it home, proving that Earth was round, that water dominated its surface, and that the Americas were not Asia. In ensuing decades and centuries, Europeans explored and colonized the eastern seaboard of the United States, Canada, Mexico, and the rest of the Americas, along with Australia, New Zealand, the islands of the South Pacific, and parts of Africa.

With their sailing prowess and gunnery, the ships of discovery and conquest would rule the oceans for almost four hundred years. Variations proliferated, as did the terminology for different rigs and sail configurations. Lateen sails fell out of favor, and new kinds of fore-and-aft triangular sails—called jibs—came into use. The varied configurations combined square rigs and fore-and-aft rigs:

- A brigantine or brig or snow was typically a two-masted vessel, square-rigged on the fore mast, with a fore-and-aft sail on the mainmast, as well as square topsails.

- By contrast, a bark or barque had three or more masts, all square-rigged except on the mizzenmast, which had a fore-and-aft sail. A few barks had as many as twelve masts.

- A barkentine was different from a bark because its two aft masts carried fore-and-aft lower sails and gaff topsails, which were quadrilateral fore-and-aft sails set on a spar or "gaff" that extended along the top of the sail. Many barks and barkentines and other vessels used long bowsprits to carry one

or more jibs. A bowsprit allowed for more sail area without
having to increase the height of the masts.

- A vessel with three or more masts that were exclusively
  square-rigged was a full rig or ship rig or technically just a
  "ship." Sometimes, additional fore-and-aft-rigged sails were
  set between the masts—these were called staysails, on top of
  which might be smaller skysails or moonrakers. Today, "ship"
  pretty much means any large ocean-going vessel.

- Then what is a "yacht"? It used to be any luxurious vessel that
  carried royalty. Then it meant any vessel used for pleasure.
  Now the definition lies in the realm of aesthetics and ego.
  Perhaps the best is that only yachts use varnish—ships and
  boats suffer the indignity of paint alone. Finally, some
  yachtsmen would tell you that the sport of yachting is
  confined to sailing. No matter how fancy a big motorboat
  may be, these yachtsmen say, the motorboat is still a
  "stinkpot," inevitably driven by a landlubber.

- Naval craft had their own names. There were the Spanish
  galleons, Royal Navy frigates, multidecked man-o'-wars and
  the exalted ships of the line, like *HMS Victory*, that were
  decisive in the Battle of Trafalgar in 1805.

Whether they were fighting ships or merchant vessels, all carried
scores of vertical ropes—lines and sheets and halyards—that deckhands
used to furl and unfurl the sails, and to rotate the yards around the mast.
The handsome intricacy of that rigging—together with the stays and
shrouds that supported the masts, as well as the series of horizontal rat-
lines forming a ladder from the deck to the yards—helped give square-
riggers their impressive look. Between the different rigs, as well as names
for the respective sails and ropes, you needed nautical dictionaries from

different time periods in assorted languages to master the lexicon and be able to identify a particular square-rigger. Perkins prided himself on his ability to do so. In their time, the square-riggers shared a special place in the hearts of royalty and commoners, painters and writers, and young and old, who adored these ships that had connected the seven seas and continents. The majesty of these ships—and the tales of their perilous voyages—conferred upon the vessels an irresistible aura.

As the industry moved into the nineteenth century, there were other sailing innovations that had nothing to do with the square rig. Schooners—employing a fore-and-aft rig exclusively, usually with two or three masts and gaff sails—became popular on the northeast coast of the United States and Canada because they could sail closer to the wind than a square-rigger and required fewer crew members. For coastal trade and getting to the Newfoundland fisheries—sailing more north and south than east and west with the trade winds—they were ideal. While schooners had been around since the 1700s, they did not become big until the 1800s—and by the turn of the twentieth century, the rigs became phenomenal. The 395-foot-long *Thomas V. Lawson* was the only seven-masted schooner ever. She was a spectacle but, according to a crew member, she "handled like a beached whale." Beside commerce, schooners in their day were also designed for yachtsmen and racing—Perkins's *Mariette* was a twentieth century schooner, if now a relic. Yet as significant and relatively inexpensive as they were, and as vivid an image as "schooner" conveyed—it probably derived from the New England "scoon," for making a flat stone skip along the surface of the water—these craft never quite became part of folklore the way the clippers did.

Schooners, too, helped lead to the popularization of the sloop, the simple fore-and-aft rig that prevails on almost all modern sailboats. In Bermuda, designers eliminated the gaff on a schooner, so that the main sails became triangular rather than four-sided. This made the rig—a "Bermuda rig" (later, a "Marconi rig")—lighter and far simpler for a crew to manage. It also gave a boat better performance while sailing toward the wind—"beating to windward," as it's often called. That was because tack-

ing back and forth, between port and starboard tacks, was easiest on a
Bermuda rig. Today, the sloop—with its single mast and triangular mainsail, together with a jib or larger headsail called a genoa that is attached to
the forestay—is nearly universal. That configuration of two sails (and
occasionally more) seemed to best create the "slot effect" that sailors
wanted—the channeling of wind between the sails that produced a kind
of suction force that helped propel the boat forward. In concert, the sails
acted efficiently, in a way they could not independent of each other.

————————

The advent of the steam engine heralded the Industrial Revolution and
changed the world. And in the nineteenth century, efficient and reliable
steam power was the death knell for the great sailing ships. Yet it would
take time to dispatch such an ancient tradition. And the irony of it was
that the rise of steam would breathe life into the last, and grandest, creation of the Golden Age of Sail: the clipper ships of the 1840s to 1870s.
In the evolution of sailing over thousands of years, clippers represented
the most significant advance in design.

These were the vessels described by one historian as "stately as a
cathedral." As culturally defining as the clippers were at their peak—in
literature, poetry, oil paintings and newspapers—they had no exact specifications. Any clipper was long, light, slender, and astonishingly fast. It
was designed entirely for speed rather than cargo space, with these characteristics: a sharply angled, concave bow that gave the ship its sweeping
look; a radically streamlined hull that was five to six times as long as its
beam (width); at least three masts with unusually long yards; and as many
as forty sails, almost all square-rigged, jutting out well beyond the sides of
a ship, forming the proverbial "clouds of canvas." A clipper sat low in the
water, so that her freeboard (the amount of hull between the waterline
and the deck) was relatively minimal. The captain invariably was a speed
demon, willing to drive crew and craft to the limit. That is why this era
was often referred to as the "days of iron men and wooden ships." In
heavy seas, the water could sweep over the deck all the way from bow to

stern. Crew, who had to climb the rigs to make repairs, as well as furl and unfurl upper sails, faced grave danger. With mammoth waves pitching the boat up and down or rolling her side to side or overwashing the deck, crew members risked falling to their deaths, drowning overboard or being injured by the sails and hardware thrashing in the wind. Reaching unprecedented speeds of twenty knots and setting some sailing records that have never been surpassed, it was no wonder the vessels were called "greyhounds of the sea." By contrast, the average merchant ship of the time barely hoped to exceed five knots.

Speed was a new fashion, though one that would stick. It came first to land transportation in the form of the American stagecoach and railroads. And then it came to the seas. (These were the first days of "globalization"—150 years before Tom Friedman discovered the word in his writings about world market trends.) During a period of mass emigration to the United States, especially after the Gold Rush began in 1848, as well as expanding transoceanic trade in such light, high-value goods as tea and spice—facilitated by the end of Britain's commercial monopoly on Far East ports—fast ships meant big profits. A single successful voyage might pay for the entire construction cost, around $75,000, of a clipper ship. Some of the early American millionaires made their fortunes from the tea trade. Anyone with capital to spare wanted to invest in a clipper—like the Dutch tulip craze of the 1600s or the Internet days of the 1990s.

Naval architects and shipbuilders in the United States and Britain led the clipper revolution. From the Chesapeake to the Canadian border, it was American designers especially—unconstrained by centuries of Old World convention—who pushed to create the fastest sailing craft ever, the "Yankee clippers." The public was fascinated. Tens of thousands crowded shores to see the launch of a clipper. In Boston in 1853, an eyewitness observed, a ship was launched "amid the roar of artillery, the music of bands, and the cheers of the vast multitude." Shortly after gold was discovered in California, street placards ballyhooed "quick passage to El Dorado" on "magnificent" clippers that promised a quicker trip than the overland route. "Never in these United States has the brain of man conceived, or the hand of man fashioned, so perfect a thing as the clipper ship," the maritime historian Samuel Eliot Morison later wrote. "In her, the long-suppressed artistic impulse of a practical, hard-working race burst into flower." The peerless naval architect Donald McKay of East Boston achieved his status building boats for companies in both the United States and Britain. After the rush to California ended in the 1850s, he produced a fleet of British clippers that restored England's reputation as monarch of the sea.

Nautical etymologists disagree why the clipper ship was so named, but the various explanations are all pretty good: the ships clipped time off a voyage, clipped off the miles, or simply sailed at a good clip. Their speed and grace inspired their owners to give them identities that suggested adventure and power. During two generations, the names, and the imagery they conjured up, were famous: *Flying Cloud, Sea Witch, Sea Serpent, Black Prince, Flying Fish, Phantom, Westward Ho!, Nightingale, Lightning, Sovereign of the Seas, Glory of the Seas, Romance of the Sea, Crest of the Wave, Dashing Wave, Seaman's Bride, Fiery Cross, Neptune's Car, Red Rover, Sir Lancelot, Friar Tuck*—and most beloved among them, the *Cutty Sark*. There were clippers of many varieties, based on their cargo or destination: the China clippers, the California clippers, the tea clippers, the coffee clippers, the wool clippers, and the opium clippers.

All they needed was a breeze. Steamships didn't have to worry about the vagaries of wind. As long as they had coal to burn, as long as there were stokers to feed the furnace, as long as there was a navigable waterway, steamers could go where and when they wanted. That included steam-powered tugboats, which could push or pull other boats, like sailing vessels that couldn't maneuver in tight channels or make it upriver to major ports like London. Now designers no longer had to agonize over agility, so they could build bigger sailing ships and employ new technologies to improve their speed. Wire, instead of rope, could be used for the assortment of fixed lines that braced the masts. Stronger forestays, backstays, and shrouds on the sides permitted a rig to support a larger sail area, which in turn gave a ship more speed. A taller mast could do the same thing, but a mast—even made of iron or steel—could only be so tall and maintain its resilience. Bigger and faster—this was always the be-all and end-all for the designer of a sailing ship. So, as it turned out, the development of steam enabled the development of the clipper.

Yet there was another, less utilitarian reason for the genesis of the clippers: sailing as an end unto itself. While powerful steamships inexorably transformed international commerce, these mastless vessels that bellowed smoke from their funnels were hardly attractive. Based on history and aesthetics, sailing still had a hold on the imagination of designers, mariners, and those on land who might watch a boat head out to sea. As swift technological progress beckoned, sailing was nostalgic. And the clippers—their beauty relying on proportions and lines—were especially compelling. "With the wind in your face," "with the sea at your back," "with the smell of salt in the air"—was it any surprise that not everyone desired to rush to modernity? While the skippers of the clipper ships were men of commerce, in their pursuit of speed and craving for competition they were really the first yachtsmen.

The earliest U.S. clippers, under one hundred feet in length, sailed out of Chesapeake Bay around the time of the American Revolution. Designed as blockade runners, these "Baltimore clippers" weren't square-riggers at all, but two-masted schooners. Though used by the navy, and then by privateers and smugglers, the Baltimore clippers offered commercial shipowners a glimpse of the speeds that were possible. On certain European trade runs, the theory went, clippers might even hold their economic own against steamers. For some kinds of light cargo—tea and spice—a fast ship was ideal, especially when being first to port meant a substantial premium to the shipowner, along with bragging rights. The dash to the marketplace motivated architects and financiers to build the swiftest boat, all the more so because the differences in speed between clippers could be marginal. In seasonal markets such as tea, the boats in effect were in a continual race from the East Indies or the Far East back to

Britain or the United States. The competition between the "tea clippers" was frenzied, and newspapers eagerly recorded the times of the clipper runs.

By the middle of the nineteenth century, the hull design of the clippers had become exceptionally refined. This was the heyday of maritime entrepreneurship and of the so-called "extreme clippers," usually three-masted, more than several hundred of them in all. These imposing iterations of sailing technology featured bows that were more raked—or sharply angled—to knife smoothly through waves at significant speed. That architecture meant even less cargo space. Designers also were better able to calculate accurate weights for ships, and they better understood hydrodynamics—the effect of water resistance on solid objects. Typically made from wood or iron, the extreme clippers were even longer than their forebears, ranging from about 150 feet to 250 feet, which was still considerably smaller than Perkins's contemporary *Maltese Falcon*; the largest ever built, McKay's freakish *Great Republic,* at 302 feet (without a bowsprit), was slightly longer than the *Falcon.* To manage the vast sail areas on their towering square rigs, this new class of clipper carried a crew of thirty to fifty sailors.

The 192-foot *Sea Witch,* from naval architect John W. Griffiths, was the best-known of the early extreme clippers. Beginning in 1847, *Sea Witch* established a series of sailing records on the tea route between New York and China, as well as between New York and San Francisco—some of which have never been exceeded. In her third circumnavigation of the world, in 1849, *Sea Witch* made it from China to New York in under seventy-five days—a westward distance of more than 14,000 circuitous nautical miles—beating some tea-laden ships by weeks. During one part of the passage, she went 2,634 miles in ten days. The New York papers gave the *Sea Witch* captain a hero's welcome, akin to what the Mercury astronauts would receive more than a century later.

The following year, just weeks before California entered the Union as the thirty-first state, *Sea Witch* became the first ship to make it from New York to San Francisco, in fewer than one hundred days—cutting by half how long it had taken the old European ships during the Age of Exploration. The year after that, on her maiden voyage, McKay's *Flying Cloud* made the Atlantic-to-Pacific 14,000-mile run around a snowbound Cape Horn in a record-setting eighty-nine days. That same year, McKay's *Champion of the Seas*—on a passage from Liverpool to Melbourne, through the icebergs of the Southern Ocean—sailed 465 nautical miles in a twenty-four-hour period, averaging more than nineteen knots (a long-standing record of the era that took many years for even a steamship to surpass). Other ships on other runs—from London to Shanghai, New York to Chile, Portsmouth, England, to India—set comparable speed records in this glorious time for clipper ships. *Flying Cloud*'s New York-to-San Francisco time, which the vessel itself bested by eight hours in 1854, prevailed for 135 years. Some 250 boats had unsuccessfully attempted to break the record, until a Floridian in 1989 knocked nearly nine days off it in an ultra-light, space-age fifty-foot experimental craft named *Thursday's Child*.

––––––––––

As transcendent as it had been, the clipper era was short-lived. Steamships were simply more reliable and, as they developed technologically, they became more economical. For Europeans in particular, the completion of the Suez Canal in 1869 meant steamers could make it to the Far East via the Mediterranean—an inhospitable route to sailing ships because of the absence of dependable wind. The once-mighty clippers were eventually converted into slower rigs that could carry heavier cargo.* Many that survived wrecks and storms wound up as canning vessels in places like Alaska; when they rotted out, they were burned on the beach for scrap iron.

––––––––––

* Some of the literature refers to these ships as "windjammers." The term seems to mean different things to different writers, but generally a windjammer of the late nineteenth century was a very large iron or steel square-rigger that carried heavy loads, but wasn't particularly fast. It was impressive looking—often with more than three masts—but did not have the lines of a clipper and was sometimes referred to derisively. Windjammers were useful for routes where steam power was impractical—for example, where coal wasn't available along the way.

Those who believed that a clipper ship represented something more than a mode of transport saw such history not as natural evolution but humiliation.

Yet the clippers did not go quietly. And two of the last British tea clippers, *Thermopylae* and the *Cutty Sark*, would be lionized in their own time and for decades thereafter as the epitome of the Age of Sail, probably the two fastest single-hulled ships that had ever sailed. Some extreme clipper ships did well in heavy weather, others performed better in light air. By contrast, *Thermopylae* and the *Cutty Sark* excelled in all conditions. Both were "composite clippers," which used a new style of construction: while its external planking was still wood, the interior frame was wrought iron and the hull below the waterline was sheathed in copper both to reduce friction and ward off barnacles. Both were 212 feet long, both had a beam of thirty-six feet—an even greater length-to-beam ratio than the American clipper ships of the 1850s, and both were built to fly. The *Cutty Sark's* tallest mast rose 152 feet from the deck, with *Thermopylae's*, about the same.

Painted green, with distinctive lower masts and yards in white, *Thermopylae* stressed width rather than height in her rig: at up to eighty feet, some of the yards on her mainmast were more than twice the width of the boat. *Thermopylae* honored a place in eastern Greece where the Spartans engaged in a heroic defense against the Persians in 480 B.C. She was launched in June 1868, and immediately set herself apart on record runs from Britain to Australia and China. The *Cutty Sark*, launched the next year from Scotland, was created specifically as a rival to *Thermopylae*. Her shape below the waterline allowed her to sail closer to the wind. She had even wider yards; and she carried more sail area at 32,000 square feet—the most of any clipper. Her deck was made of teak, which provided additional stiffness under way. But she and *Thermopylae* were about as similar as different clippers could be. The most distinguishing part of the *Cutty Sark* was the least functional (unless you believed in superstitions of the sea, which all sailors do): the elaborate figurehead atop her bow. Ornamental carvings were supposed to bring good luck to crew and ship; a figurehead, in particular, gave a ship eyes with which to find its way. The figurehead adorning the *Cutty Sark* was a maiden wearing the revealing dress described in the Robert Burns poem "Tam O' Shanter" as a "cutty sark."

It was said at the time that no ship, under sail or steam, could stay with the *Cutty Sark* in a stiff breeze. Those who had sailed on or near her remarked that they had never seen another vessel—even a "smoke stack," as the steamers were sometimes called—pass her by. She was known for incredible bursts of speed when running with tailwinds in a storm, to the point where some helmsmen became afraid to let her go fully. All the seafaring world wanted to see her duel *Thermopylae* on the tea run back to London from Shanghai. The British loved to race, all the more so at sea. This could be "a race in which every man in any way connected with, or interested in shipping, was as much concerned as are the horsy fraternity in the Derby or the rowing fraternity in the Oxford and Cambridge boat race," wrote Sir Basil Lubbock in "The Log of the Cutty Sark" half a century later. Both *Thermopylae* and the *Cutty Sark* liked to boast their ship was fastest, but it would take a match race to decide who was right. As Lubbock neatly put it, "For though the men of the sea have the most acute memories regarding their ships, in the matter of speed they are as prone to exaggeration as a fisherman when giving the weight of his fish."

These were not recreational vessels, so it was not as if their owners or captains could set up a race months in advance. Much depended on fortuity: if one of the clippers was ready to go, she wasn't about to wait around for a competitor with the same commodity. But on June 18, 1872, both *Thermopylae* and the *Cutty Sark* were in Shanghai, fully loaded with their cargoes of tea, more than a million pounds of it. Everyone knew the race was on, and it caused a sensation. Huge sums were wagered on which boat would get to London first; some crew members bet their entire pay. The "bully captains," mad and otherwise, had even more on the line: a lifelong reputation could be won or lost based on who won what the papers called "the Great Ship Race from China." And for the shipowners, it meant everything. The *Cutty Sark's* owner, John Willis—"Willis of the White Hat"—had staked his company on his vessel prevailing.

For nearly two months, the clippers went at each other, with the *Cutty Sark* in the lead most of the way. At one point, in the Sunda Strait between Sumatra and Java, *Thermopylae* led by a single mile. But by the time they reached the southern waters of the Indian Ocean, the *Cutty Sark* was in front by four hundred miles. Then came disaster: during a catastrophic gale, the *Cutty Sark* lost her rudder. While the crew was able to fit a replacement, she lost five days and had to proceed at reduced speed around the tip of Africa and homeward to London. Nonetheless, the *Cutty Sark* arrived only a week after *Thermopylae*, which made the voyage in 115 days. The *Cutty Sark* crew members were celebrated as hardy, courageous heroes—rock stars of a different age. Because of her bad luck and because she had been so far out in front, the consensus in the shipping world, if not in the betting arena, was that the *Cutty Sark* had won the race.

In subsequent years, during the 1870s and '80s, the two antagonists competed less directly, on runs to or from Sydney. But their respective trips didn't commence simultaneously, so the *Cutty Sark's* superior times didn't count as such. Still, it was her legend that continued to grow. In 1890, the now unprofitable *Thermopylae* was sold to a Canadian concern that converted her into a lumber- and coal-carrying hulk; in turn, she was sold to the Portuguese navy, which took away her vaunted name and turned her into a training ship. In 1907, in a final, sad moment, the *Pedro Nunes* was deliberately sunk by torpedoes during target practice.

By contrast, the *Cutty Sark* lived on. She, became an Australian wool clipper, then was unceremoniously sold in 1895 to Portuguese merchants who named her *Ferreira*. She would have suffered a fate similar to *Thermopylae*—every other extreme clipper did—but an Englishman named Wilfred Dowman discovered her by accident in 1922 as she was forced into Falmouth Harbor by a squall in the Channel. Dowman, an old sea captain, bought her, then restored her hull, name, and nationality. Now the old *Cutty Sark* was home for good. She became the property of the Thames Nautical Training College (as well as the inspiration for a brand of Scotch). In 1953 she was taken over by the Cutty Sark Preservation Society under the patronage of Prince Philip, the husband of Elizabeth II; the clipper was finally moved the next year to a specially constructed dry dock in Greenwich. There she was fully rigged and rechristened by the Queen—signifying a bond between modernity and a venerated sailing tradition. Revered by her countrymen, sanctified by history, today the *Cutty Sark* is a national monument and popular London museum—and a testament to the emotional grip that the magisterial clipper ships still have. In the spring of 2007, she suffered a catastrophic fire—and her admirers around the world rallied with donations and pleas to save her.

Perkins wanted his boat to write the next chapter of sailing history. After the steamers arrived and after the last commercial sailing ships were retired, sailboats were confined to recreation and sporting—albeit often on a grand scale. Sailing for pleasure dated to Roman times. Yachting clubs had become popular in Britain in the nineteenth century. The Royal Yacht Squadron in 1851 hosted a race around the Isle of Wight that began the America's Cup series. The rival New York Yacht Club was founded in 1844. Summer regattas, to such venues as Newport, Rhode Island, and the cool waters off Maine, were a pinnacle of the social scene.

Meanwhile, sea adventurers in the tradition of Columbus and Magellan attempted solo voyages and other feats. After World War I, sailing became more popular at different ends of the social and economic spectrums. There were modest racing dinghies and other day boats that ordinary folk could use. And then there were the handful of sleek, extravagant J-class racing yachts of the 1930s—exceeding 120 feet in length and each costing up to $500,000—built for the gentlemanly likes of Harold S.

Vanderbilt, Sir T.O.M. Sopwith and Sir Thomas Lipton. A few of the fore-and-aft-rigged J-boats still exist—*Endeavour* was the most excellent trapping for CEO Dennis Kozlowski, of Tyco International infamy—and Perkins considered trying to purchase and restore one of them.

But it was the clippers that Perkins admired most. He had long collected books, models, maps, and prints of the age. He loved the marvels of engineering that they represented in their time. He loved the hubris of their designers, the guts of their skippers, the beauty of their lines, and the national pride they instilled. In the evolutionary tree of sailing ships, there had not been much innovation since their time. Perkins wanted his own ultimate sailboat to be a reincarnation of the clipper ship, harnessing twenty-first century materials. As much as Perkins respected novel technology—and had nicely profited from it as a venture capitalist—he adored the clippers. McKay was a genius at naval architecture, but he had no computers; Griffiths obviously never heard of carbon fiber. What a clipper might be in the new age!

When the *Maltese Falcon* was conceived—as she made it from blueprint to supercomputer to shipyard—Perkins always had in mind these ships of old and that he might someday challenge their records. Most of all, he loved the tale of the *Cutty Sark* and *Thermopylae*. Did he have a favorite between them? The *Cutty Sark* wasn't far away from one of his second homes—a manor house in East Sussex, outside London. He knew the ship well, like so many others who lived in England. One might have assumed that a tangible, living clipper would capture his heart. But it was *Thermopylae* that he appreciated more. After all, in the Great Tea Race of 1872, *Thermopylae* was the victor.

But was not the *Cutty Sark* at least the more worthy boat, the vessel and crew with the moral victory? Isn't that perhaps why the *Cutty Sark* still reigns and *Thermopylae* sits deep beneath the seas she had once sped across? Perkins scoffed at the notion. "Winning is winning," he said. "What matters is who comes in first." It was a recurrent theme—in the context of sailing, sports, people, human relations, business, and venture capital, the latter of course being his specialty, his success at which allowed him to create such things as a $130-million modern clipper yacht. "Winning is winning," he said. "The rest doesn't matter." Danielle Steel learned that just playing "Liar's Dice" with him. The after-dinner game, on a boat or in the living room, was a version of poker played with dice, a combination of luck and bluffing prowess. When Perkins was ahead at the game, they could play for hours. When Steel was winning, he seemed to grow very tired very quickly, and turned in for bed. Perkins was a good winner—he wasn't as good at losing.

THREE

Westward Bound

Tom Perkins learned to love the water early on—because he had to. When he was barely two, a relative had forced him to learn how to swim. Her teaching method, as Perkins liked to tell it dramatically, was to throw him in the water and say, "Swim! Or you'll drown." The venture capitalist's credo indeed: sink or swim. And, with a certain bravado, he learned how to get some mileage out of the swimming story. In his telling, folks came from all around to see the "miraculous swimming baby." Maybe it was even true.

He grew up in the northern outskirts of suburban New York City, in a small city named White Plains—an only child, lonely in a cautious, tightly wound middle-class family. Perkins said his parents were hardly enthralled by his presence. His mother Elizabeth periodically threatened suicide, or at least acted out her frustrations by pressing a cleaver against her wrist. When Perkins didn't live up to expectations, his father Harry asked his mother why they'd had young Tom in the first place; when Tom failed to be the baseball player that his six-foot-six dad was, his father declared that he "swung like a girl." Money was as scarce as affection. Both his parents had struggled during the Depression and into the 1940s—when Perkins was a teenager—and the family agonized over finances constantly. His father toiled away at a desk job for a Midwestern insurance company, while his mother was an occasional seamstress. Perkins said he could never understand his father's refusal to look for a job that would make him happier and wealthier.

To help with the family balance sheet, Perkins had a paper route, mowed lawns, and shoveled snow. For entertainment, he listened to lots of radio and collected wind-up model boats—and got his first taste of real sailing as crew for a friend on a seventeen-foot sloop they raced on Long Island Sound during summers. This, Perkins said, was "back when Long Island Sound still had wind." (It's a funny line and Perkins intends it that way. But ever the nerd, he also explained that sixty years ago Long Island wasn't heavily built up with highways and office parks and housing developments. That meant the afternoon southerly breezes off the Atlantic Ocean made it up to Long Island Sound, whereas today the summer heat off all the concrete prevents the winds from making it that far north.)

Some people look back on their modest upbringings nostalgically. Perkins did so with a grimace and his trademark sigh. Many decades later, with several chefs in his employ, he could still recount family Friday night dinners of baked salmon loaf, washed down with lime Jell-O. Needless to say, that is not what's served on the *Maltese Falcon*. What animates individuals to seek great wealth? The seeds of desire are not always so easy to find. In Perkins's case, though, it's classic: he didn't want to relive his childhood. "Nobody in my family had any money and mother obsessed over it," he said.

So, too, it is easy to see how he became a self-described gearhead. Since he wasn't good enough for most high-school sports and since he was an excellent student, the path of academics was obvious. Perkins remembered his physics teacher—"Mr. Wilson," an overweight public-school lifer who looked like Broderick Crawford and sounded like William Conrad—as the first person to take him seriously about anything. Perkins loved science and tinkering, and used his spare time and money to make high-voltage Tesla coils and other mad-scientist devices within a teenager's budget. In the neighborhood, he built television sets out of kits and sold them to friends. His ambition was to finish high school and become a TV repairman. But Mr. Wilson—who had been one of Perkins's TV customers—had different plans for his prized student. He wanted Perkins to apply to college—something that his parents had neither saved for nor contemplated. With Mr. Wilson's help, Perkins was accepted at Harvard and the Massachusetts Institute of Technology. Perkins chose MIT because it gave him a better scholarship. His father chipped in two hundred dollars, on the condition his son returned the loan. On graduation day in 1953, Perkins did.

———————

At MIT—an institution filled with science-kit, experiment-happy geeks—Perkins found that he wasn't quite the athletic misfit that his father imagined. The early near-drownings paid off. He wound up making the swim team and broke the MIT pool record for the one hundred-yard freestyle. He also skippered a sailboat for the first time—in MIT's twelve-foot, single-sail Tech Dinghies, designed for the school by alumnus Nat Herreshoff himself. A possible career in TV repair gave way to the dream of becoming a theoretical physicist, but Perkins did not have the gift and settled for a degree in "electronic engineering" (what is today electrical engineering, which has been mostly merged into computer science). Upon graduating, he didn't have to serve in the Korean War because of a heart murmur and he went to work for Sperry Gyroscope for two years to pay off student debt. His job was to maintain Sperry's radar-controlled guidance systems for NATO jets. The location was auspicious—in the deserts of Turkey, a country where half a century later he would build the *Maltese Falcon*.

With the assistance of government grants and burgeoning ambition, in 1955 Perkins was off to Harvard Business School, where his hands-on engineering talents merged with the skills of starting companies—thereby forming the combination that would beget a venture capitalist. Just as with Mr. Wilson in high school, Perkins found the right teacher, coming into the orbit of Georges Doriot, a star of the business school. A former brigadier general in the U.S. Army, Doriot was master of both entrepreneurial theory and real-world application: with his publicly traded company, American Research and Development, he was considered one of the first U.S. venture capitalists. He was a character as well. He liked to invite students to his study and play French military songs to inspire them. In trying to get students to think like executives, he offered comparisons like this, in his imperfect English: "An engineer get drunk on *bee-ya*. A chief financial officer only drink Scotch!" At Harvard, he taught a famous second-year lecture course called "Manufacturing." It was more about entrepreneurship and even more about Doriot's view of the universe. Perkins remembered the first class, in which the lesson of the day was how to read a newspaper. Doriot asked the roomful of students what was the most important section of the paper. Stock tables, business headlines, editorials? No, it was the obituaries, suggested Perkins. He was right, said the general: that was the only section in which journalists couldn't lie.

Students in Doriot's class were broken up into teams that had to find a Boston company to study. While most chose straightforward manufacturing concerns that had vice presidents all wearing repp ties, Perkins and his teammates picked something approximating a pajama factory. At the end of the term, Doriot hosted a cocktail party for the students and the executives at the companies that had been studied. The students pushed for Doriot to meet "their" executives. Perkins and his team watched gleefully as Doriot and the CEO of their clothing company spent most of the party together—they had been buddies during World War II. The big prize of the class every spring was being asked by Doriot to return to Harvard after graduation for a year as his assistant. Perkins got the offer. "All these Type A personalities were maneuvering to get the honor," said Dave Thompson, one of Perkins's teammates who became a lifelong friend. "Tom didn't spend his time trying to pull together the smartest students in the room. But he had it figured out."

Perkins turned down the Doriot plum. He had something else in mind—and certainly something that cut against everything that his risk-averse parents stood for. He was off to a place he had never been to, that he'd barely thought about, where he knew no one. He was off to the West, never again to set foot in White Plains.

In the summer of 1956, between his two years at the business school, Perkins worked for General Radio, near Harvard. General Radio made various testing instruments, the kinds of devices the public later associated with Hewlett-Packard, the company founded in 1939 that spawned Silicon Valley. Perkins's task was to run quality-control checks on HP's products—as a way of assessing the competition, which happened to be less expensive. At that point, HP was still a relatively small company with $25 million in annual revenues—it didn't take itself public until the following year—but General Radio was taking notice. "That was the first time I'd heard of HP," Perkins said.

Though he had assumed that summer that he'd go to General Radio upon graduating, Perkins was intrigued by the two upstarts behind HP that were challenging his employer. He was curious enough that when he heard that Bill Hewlett and Dave Packard would be at an electronics trade show at the New York Armory, he went. "There they were, setting up their own booth," Perkins recalled. He helped them, and talked to Packard for most of the afternoon. That got him a job offer. Packard became Perkins's great mentor for the next two decades. "I had little to lose and figured I'd try California for a few years before returning to the East Coast," Perkins said. He was impressed not just with the intellect and aspirations of HP's founders, but their "down-to-earthiness" and their skepticism about his fancy MBA in the making.

It was only after he arrived at the company in 1957—after a leisurely drive across the country with a girlfriend in his rundown Jaguar (Perkins changed the water pump himself three times en route)—that he realized just how dubious they were. This was years before Silicon Valley was even called that—the area was mostly still apricot orchards. Despite HP's one thousand employees—the most in Palo Alto but still dwarfed by companies such as Lockheed in San Jose—an MBA graduate from Harvard was like a creature from Mars. "Typically at HP, you worked as an engineer first and then proceeded into management," Perkins said. "The idea that an MBA would go directly into management was totally strange. So they came to me and said, 'If you don't mind, we're going to put you to work in the machine shop on a lathe.'" It was not a promising beginning for an electrical engineer from MIT, let alone one with a Harvard MBA.

Single and charming—a Harvard alumni magazine described him as a cross between Fess Parker and Gregory Peck—twenty-five-year-old Perkins spent weekends in San Francisco surveying the social scene. During his first summer there, he dated Ellen Davies, part of one of the city's upper-crust families—her father was Ralph K. Davies, the oil baron. At a dinner one evening with only her mother (not including the servants), Perkins was told, "So, Ellen tells me you're a machinist. You have to understand we're of a certain position here." Perkins got the message: his grimy fingernails didn't mix well with blue blood. He didn't see the daughter again. If only Mother had known the riches that awaited him.

After Perkins spent several months in the machine shop, HP put him in charge of all independent salespeople. In time, though, he got restless, deciding that even HP wasn't growing fast enough for his tastes and that he wasn't learning enough. For much of the next six years, he seemed to bounce around jobs, none a complete bust but none quite a smashing success. In retrospect, his wanderings were not so much a lack of direction but a certain conviction that he knew best. He quit HP to work for a national consulting firm in San Francisco for two years, then left for Optics Technology, a new fiber-optics start-up in the Bay Area that Packard and Hewlett themselves had largely funded and asked Perkins to manage. Fiber optics was one of the hot new technologies of the 1950s.

The founder and CEO of Optics Technology was Narinder Kapany, a magnetic, gifted immigrant from northern India who had proven with fiber optics that light need not travel in a straight line. It was he who coined the phrase "fiber optics" in a series of professional papers. Perkins was brought in to create business opportunities for the nascent company, while Kapany the physicist handled the genius aspects. Unfortunately for

both, they got along infamously—"a mutual hate of each other, of bibli-
cal proportion," as Perkins put it. "I hate him to this minute." When
Kapany's name was mentioned by anybody, Perkins liked to crack that he
wanted his own gravestone to read, "I *still* hate him."

For the next four years at Optics Technology, Perkins and Kapany
went at each other. It was a match made in Armageddon. Some of their
differences were personal—the rest were philosophical. That pretty much
covered the workday. Kapany wanted government contracts to fund
research. Perkins wanted the company to make its own products and
thereby control all the profits. In truth, Optics Technology was going to
have a tough time regardless. Like many technological innovations, fiber
optics were just too costly at the time to have wide commercial application. However, in the course of his sales calls for the company, Perkins
came to learn about lasers. This, too, was a key emerging technology—
once theorized by Einstein and now hailed "as the wonder device that
would soon be doing everything from shooting down enemy missiles to
curing cancer," according to the *New York Times*. It was so new that many
papers continued to identify it without an acronym: light amplification
by stimulated emission of radiation. The first working laser was demonstrated only in 1960. The first laser Perkins saw was back at MIT and it
was so fussy that the most minute vibration interfered with the dazzling
red beam.

Perkins had the notion that a less temperamental, less expensive
version of the MIT laser might be a gold mine. At Optics Technology, he
helped to develop such a device and the company sold many of them, for
several thousand dollars apiece. Still, the company couldn't capitalize on
Kapany's various research projects—a fiber-optics tube for physicians to
use in the throat, a machine that produced color X-ray images—and Perkins concluded that he or Kapany had to go. The board of directors sided
with Kapany. After all, he was the public face of the company—"the
father of fiber optics," as the press and scientific community referred to
him. Perkins was gone, never to talk to Kapany again—"the only man on
this planet I don't speak to," Perkins liked to boast about his "grudge of all

grudges." Packard, who had urged Perkins to get out of the company by now, asked him to return to HP to help organize a central R&D department, which included the new fast-growing computer division.

Perkins agreed and in 1966 began a second stint at HP. There was one caveat: his deal with Packard had to allow him to moonlight on his own commercial ventures. He dabbled with ideas of a low-priced electronic voltmeter and a product that made holograms, but he settled on trying to build a smaller, cheaper laser than what he'd come up with at Narinder Kapany's company. The idea would make Perkins a multimillionaire before he was thirty-five—and it allowed him to give the financial finger to Kapany. If you ask Perkins which of the two was more satisfying—getting rich or getting even—he'd just chortle. He liked both—and when both could be accomplished in the same business deal—so much the better. It was the same competitive impulse that drove him in sailboat racing. He loved to win—and he loved for anybody he disliked to lose.

And then there was the money. Life at HP was perfectly adequate and his career track there couldn't have been better. But being a salaryman wasn't for Perkins. He remembered an HP friend and vice president once showing him a retirement-plan financial table. "That's you there," the friend eagerly pointed out. "You have this much stock and if it goes up this much, you'll have a million dollars by the time you retire." Though Perkins replied, "Great!" he was thinking, "To hell with that—I want a million dollars now!"

———————

Bill Gates, Steve Jobs, virtually any of the modern Siliconillionaires—they'll all tell you they weren't in it for the money. Theirs was the self-professed joy of bettering mankind and changing the world. Tom Perkins was in it for the money from the beginning. "I wanted to be a millionaire by the time I was thirty," he said—and it was something he had often boasted to Gerd. "I didn't make it. So I figured that, goddamnit, I was going to make it, and a lot more, by the time I was forty."

To make his new "helium-neon continuous gas laser" apparatus, Perkins needed to figure out how to put mirrors inside a laser tube; in other variations, the mirrors were outside the laser tube and subject to movement. Such a simpler self-contained design was akin to a light bulb, in need of no ongoing adjustment. He found a glassblower—a friend from sailing circles—to make the correct tube. In return, Perkins, as chairman, promised him a percentage of the new laser company that Perkins had in mind. Perkins was becoming a venture capitalist without knowing it. The slight wrinkle was that he wouldn't be investing other people's money—he'd be investing his own.

By this time, he had married Gerd. They had met in 1959 on a winter morning in the Sierra Nevada. He was skiing one Sunday—the only day he took off—with a San Francisco friend named Buzz Kramer. Perkins noticed a tall, athletic woman ahead in the lift line. "She was the most beautiful woman I'd ever seen," Perkins recalled, tears in his eyes, years after she had died. "She looked like Ingrid Bergman—and did so her entire life." There was indeed a striking similarity to the iconic Swedish actress, which led one of Perkins's friends to call Gerd a "Nordic goddess." In addition to her stature, Gerd had wide blue eyes, blonde hair, high cheekbones, a broad forehead, and a luminous smile. "It was," he remembered, "love at first sight, truly."

Kramer saw that Perkins was smitten. "I actually know her," Kramer said, explaining they'd met at a party a few weeks earlier. "Would you like me to introduce you to her?" Perkins characteristically seemed to be in the right place at the right time—with the right personnel along.

Perkins got his introduction. And as he learned, Gerd was not only attractive, but independent and confident. She had been in the country only a few weeks—working temporarily as a secretary in San Francisco, eager to explore an entire new culture away from her native Norway. While her father was a leading lawyer in that country and she had grown up comfortably, the seeds of adventure had been planted early. Part of her childhood took place during the Nazi occupation of Norway in World War II. Before she was a teenager, she and friends were skiing between villages, delivering messages between partisan cells.

Upon their return to the Bay Area, Perkins and Gerd began dating immediately and, long before it was fashionable to do so, decided to live together before marriage. Their wedding was in 1961 and a few years later they had a son named Tor. The only money they had saved—$15,000—was for a down payment to buy the little brick house they were renting in Palo Alto. With Gerd's blessing, Perkins used it all to fund his laser idea. She was no fool. She understood the risk he was taking, but out of love or absolute confidence—or both—she was willing to take the chance. "Of course you should use the money," Gerd told him. It was the first of many times that Perkins's risk-taking psychology showed itself: when the circumstances called for it, he was willing to push his pile of chips to the center of the table. It surely helped that the woman he loved didn't question it.

With his glassblower and a former Optics Technology colleague, Perkins set up shop in the East Bay, on University Avenue in Berkeley. It wasn't exactly HP. The loft, which had served as the studio for his glassblower, was shared by two other entrepreneurs, sort of. One made life-sized, psychedelic-colored papier-mâché gorillas. The other was the renegade LSD chemist named Augustus Owsley Stanley III, who Tom Wolfe profiled in *The Electric Kool-Aid Acid Test*, was John Lennon's dealer, and inspired a Grateful Dead song. Perkins named the company University Laboratories, after the avenue and because he thought his customers would be from academia. He went to work there every evening after putting in a full day at HP.

After six months of modifications and $150,000 from investors, the laser invention worked—a "eureka moment," in Perkins's words. The invention was waterproof, vibration-proof, and easy to mass-produce in comparison to existing lasers. He named it the Lasertron. It initially sold for $195 to $295, depending on the casing. Every college physics department wanted one. Better yet, the device had industrial applications—for example, in the construction of sewer lines. Excavators no longer had to survey every inch of a pipe's path to ensure a straight line—all they needed was an easy-to-carry, durable laser. Similarly, vertical lasers could be used in the construction of elevator shafts. The company's hundreds of early orders came from a simple black-and-white advertisement Perkins placed in *Research/Development Magazine*, circa 1967. It showed the device and included a clip-along-the-dotted-lines mail-in coupon: "2 percent discount, net 30 days, money-back guarantee." A laminated copy of that ad still hangs in the study of Perkins's palatial Marin County home, right beside his models of vintage ships. The company eventually produced an industrial-grade version—mounted in fancy-looking housing and painted yellow—and jacked up the price twenty-five times, to $5,000. The profits rolled in.

Kapany wasn't happy about it. His company sued University Laboratories, claiming it had made off with his customer lists. He also declared, incredibly, that Perkins had stolen his laser wavelength—as if the frequency of a helium atom belonged to him and not the law of nature. The lawsuit was settled quickly: Kapany got some spending money. Perkins's start-up picked up steam. About the only problem it faced was getting enough bank loans to continue to expand. At one point, the local branch manager of Bank of America called to say there would be no more loans until Perkins and his two partners put more of their own money in. Perkins called his bank's director of public relations and tried to plead his case, using all the analytical skills a Harvard business degree conferred. The bank man didn't care. All he wanted to talk about was his daughter's project for her upcoming school science fair. "And how many will you need?" Perkins asked. The following week, the branch manager called to say the bank would lend Perkins all the money he wanted.

Perkins marked his "eureka moment"—two in the morning on July 7, 1966—the way Orville and Wilbur may have remembered the day they could fly. He recounted its details on the day nearly forty years later that the *Maltese Falcon* first touched water. There was a grace note to that July 7 moment. When he arrived back in Palo Alto near dawn, Perkins found Gerd in bed but still awake. She was happy to hear about her husband's triumph, but she had a pressing matter: she was in labor. The next day, Perkins's second child, Elizabeth, was born. "You keep talking about your laser," Gerd teased him. "Look at what *I* made!" Perkins said Gerd had always felt he'd been more excited about the laser than the baby—and, he added, "she may have been right."

University Laboratories was a big hit. While Perkins considered a public stock offering that would have made him wealthy down the road, it would've required him to continue to run the company. "I try not to get that emotionally attached to anything," he said, in a remarkable self-insight. That seemed to include not only many people, but corporate entities as well. So he took the quicker, surer deal and in 1970 sold the company to Spectra Physics, the dominant player in lasers at the time and the first laser company to go public. In many ways, it was an uncharacteristically cautious move on his part—precisely the opposite of the risky turn he took in the first place in betting his house money. But the move also allowed him to stay at HP, which he was not quite ready to leave for good, largely out of loyalty to Packard. Perkins came away with just under $2 million—"my first home run," in the parlance of venture capital. Overnight, he went from middle-class existence to sudden wealth. And he got himself a far finer house—high on a bluff, overlooking San Francisco Bay—than his $15,000 dollars in life savings could ever have purchased.

———

Meantime, at his real job at HP, Perkins was given the unpleasant task of jump-starting the company's recent entrance into what were then called "minicomputers"—processors that were smaller than the large IBM mainframes that took up an entire room. It was a political morass, affecting the sales force and the institutional morality known as the HP Way. Packard had never been interested in discounting prices as a way to build market share and, Perkins's protests notwithstanding, he wasn't about to start doing so simply because of a new product. Packard believed that if a company had a good product, market share would take care of itself. "The problem," Perkins said, "was that too many at the company thought of themselves as engineers rather than as merchants." Perkins wanted the salesmen to concentrate on selling stuff. In one critical meeting, when Packard and Perkins faced off on their disagreement, Packard got so mad at the concept of discounts that he walked out. "What now?" Hewlett asked Perkins. "I need more ammo to make my case," Perkins replied. While Packard may have been the type of father that Perkins wished he'd had—despite Packard's volcanic temper, even with petty things like getting lost in the hills of Marin on the way to Perkins's house—Perkins had no problem standing up to him. Perhaps it took one to know one. Perkins said years later that Packard was "one of three men who changed his life," along with Mr. Wilson and General Doriot.

More intense than others at HP, Perkins instilled a competitive ethos. He taught everybody his cardinal rule of the sales department: "Never let the phone ring three times." If it ever did, the sales reps were in trouble. The way Perkins let you know was whenever he heard two rings, he'd start counting, then leap from desk to desk—while he continued counting—until he got to the phone that was ringing. "If he got to the phone before you did, you got reamed out," recalled one salesperson. "Pretty soon, the three-ring rule became ingrained." Even now, Perkins's old HP colleagues liked to say they bound up from the dinner table to get the phone. The computer business took off—growing, at times, 30 to 40 percent per quarter. It might've done just fine on its own, but the company recognized Perkins's role. When Packard left for Washington to be President Nixon's deputy defense secretary for two years, Perkins moved into the executive suite to be Hewlett's right-hand man.

Because of Spectra Physics (today part of the Newport Corporation), Perkins had enough wealth that he didn't have to work. And he had been quick to let his middlebrow HP colleagues know it. He bought a nice Ferrari that he paraded into the Hewlett-Packard parking lot, which was otherwise the exclusive bastion of Chevys. When he tired of office politics and the culture clash—and after Packard was back—Perkins quit HP again. "I had learned that I was a team player only when I was the captain," he said. He could have played the game for five or ten years more, and may well have become HP's president, but his sights were elsewhere.

And as much as he worshipped Packard, Perkins wanted to break away from him finally. Perkins kept thinking about his laser vision that turned out so correctly: he'd gotten a charge out of founding a company and wondered if he could create more of them. It was one thing to midwife a single company, or even two. Perkins aimed to make a habit of it—perhaps launching entire industries—getting his fix over and over, without having to put all his eggs in a single basket. The money wasn't bad either.

———

Across town, in the new Silicon Valley, another pioneering entrepreneur was wondering the identical thing about serial entrepreneurship, going through the same frustrations as Perkins. Eugene Kleiner—an industrial engineer and toolmaker who had fled Vienna before World War II—was already well-known, having been involved in a seminal Valley start-up. In 1956, he left a safe job at Western Electric in New Jersey to come to northern California to join Shockley Semiconductor Laboratory. Whereas Perkins had zipped across the continent in a Jag, the more sedate, unassuming Kleiner made the trip on Route 66 in a Buick convertible. The new company's founder was William Shockley, co-inventor of the transistor, who would soon receive the Nobel Prize in physics for the invention. Shockley was the man who "brought silicon to Silicon Valley," as authors Lillian Hoddeson and Michael Riordan put it. But as a manager, Shockley was a train wreck. Kleiner couldn't believe it when Shockley asked him how to find a secretary and where to buy pencils. One of Shockley's odd ideas was to post salaries on a bulletin board. This managed to infuriate both those who were being short-changed and those who were doing well enough that they didn't want others to know. Between his flip-flop decision making and disrespect for people, he had no idea about running a company. It was like Dilbert's boss on crystal meth. (Shockley never succeeded in business, and went on to teach at Stanford, but even there faded into disgrace with his fixation on eugenics.)

The following year, Kleiner—along with seven other Shockley employees, including Robert Noyce and Gordon Moore—quit. Branded "the Traitorous Eight" by Shockley, and with $1.5 million in seed money from East Coast industrialist Sherman Mills Fairchild, they founded Fairchild Semiconductor—the first spin-off company among thousands that would form the replicating DNA of Silicon Valley. Fairchild Semiconductor came up with a way to manufacture multiple transistors on a single wafer of silicon. Eleven years later, Noyce and Moore mutinied to begin NM Electronics, the new chip on the block. Luckily, they later came up with a better name: Intel, the most valuable company in the Valley before the Internet. Kleiner stayed at Fairchild for six years, leaving because young Andy Grove had joined the company and Kleiner wouldn't work for him. (Grove later went on to Intel, becoming chairman and CEO.) "Too intense for me—even then," Kleiner said.

Kleiner left to launch a business that manufactured "teaching machines," an early form of interactive learning in which students had to push buttons on a gadget linked to questions asked by a teacher. Kleiner called his company Edex, for "educational excellence." And while it didn't represent an electronics breakthrough like the transistor, Edex racked up several million dollars in profits. In 1965, he sold out for $5 million to Raytheon, the old East Coast defense contractor. Kleiner stayed on for a while, technically as an employee in weapons research. "I was supposed to be selling to the education market," he remembered, "and here my business card said, 'Missiles Division.'" That ended most sales conversations.

Worse for Kleiner, Raytheon wanted to transfer him to Michigan City. Kleiner, whose thick Austrian accent and European manners masked a sense of humor, thought it was absurd. "You just don't move from Palo Alto, California, to Michigan City, Indiana," was his reaction. "Michigan City is best-known for having a penitentiary." So he quit and, financially secure, took his family abroad for two months. With time to reflect, and the recognition that he'd been in the business of starting companies for half his adult life, Kleiner decided he might want to do it for a living. He would institutionalize what the Traitorous Eight did on a whim. Until the day he died in 2003, Kleiner kept a framed photograph of the Eight, signed by all of them, in the study of his home. It was a reminder of his entrepreneurial roots, akin to the laminated laser ad that Perkins still had on his wall.

Between the money he made at Fairchild and the windfall he reaped as an initial $100,000 investor in Intel, Kleiner had lots of capital to play with. He loved not being tied down to a particular company. But he didn't like being professionally isolated. In 1972, Tommy Davis, another Bay Area investor, came by to see him. Together with Arthur Rock (who had put together the Intel deal), Davis had raised a pool of $5 million from other wealthy individuals, which they used to fund various start-ups. Davis had been approached by an East Coast scion interested in getting a piece of the California action. Henry Lea Hillman, fifty-three, was heir to the fortune of one of the last of the robber barons, J. H. "Hart" Hillman, the Pittsburgh coal-and-gas industrialist. Henry Hillman ran the business, worth $3 billion even back then and one of the ten largest family fortunes in the United States. Hillman had boatloads of cash and shrewdly wanted to redeploy it out of the old smokestack industries and into the new world of electronics. The semiconductors of Silicon Valley were the steel and coke of his father's era. Hillman also wanted to invest quietly— he hated publicity and lived by the motto: "The whale only gets harpooned when he spouts."

Davis wasn't looking for other investors when Hillman called. But Kleiner was, and Davis recommended him. Hillman offered Kleiner a job in Pittsburgh, which he didn't think was any better than Michigan City. Kleiner declined. "What if you can stay in California?" Hillman asked. "I'll give you $4 million to invest, if you can raise another $4 million."

That was a big chunk of change in those days—not for Hillman, but for the still-embryonic venture community and its investors who were made up, in effect, of financial prima donnas. To raise that kind of money, Kleiner sought out his friend Sandy Robertson, one of the early investment bankers for the Valley. Robertson—who knew that more venture deals would produce more companies that might go public that he could then underwrite—agreed to help Kleiner. Robertson suggested he meet another entrepreneur-turned-financier looking for capital. "I may not be able to raise the money for you separately," Robertson told Kleiner. "But I might be able to raise it for the two of you together." The other person, of course, was Tom Perkins. At about the same time, Perkins, too, had turned to Robertson for advice. "It wasn't until then that I realized that Gene Kleiner was going through exactly what I was," Perkins said.

On a summer morning in 1972, Kleiner and Perkins finally met for breakfast at Rickey's in Palo Alto. Different in temperament as well as technical expertise, their styles seemed to mesh, as in other signature collaborations of Silicon Valley: Hewlett and Packard, Steve Jobs and Steve Wozniak at Apple Computer, Sergey Brin and Larry Page at Google. Kleiner was nine years older than Perkins and had a formality antithetical to the Valley. Perkins had flair and supreme confidence in his own ability, leading one of his venture-capital colleagues to observe that he probably was self-assured in his birth crib. Kleiner knew manufacturing; Perkins understood hardware and management. Kleiner had modest tastes; Perkins, even before becoming a yachtsman, was a fan of 1930s sports cars, at one point owning the world's best collection of antique roadsters, including a Duesenberg once built for a maharajah and a pre-war Bugatti Atlantic coupe that Ralph Lauren now owns. Without intending wit, Kleiner referred to the classics as Perkins's "used cars." Perkins had people that tended to the cars, but maintained a machine shop near his house so he could get his hands dirty on them, too. (Ever a man of commerce, who believed all of life was transactional, Perkins sold the cars when he feared the market had peaked in the late 1980s. Naturally, he made millions from these deals, too.) In the venture business, Perkins became the Valley's supercharged engine, generating great ideas. But it was Kleiner who provided reflection and skepticism.

The two of them hit it off and spent ten hours together for each of the next few days. Thus was born Kleiner & Perkins. The name was arranged so for two reasons—alphabetical and, as Kleiner noted, "I had the commitment for $4 million from Henry Hillman." With the additional $4 million they had to raise, the $8.1 million fund they planned was unheard of in those early days of venture capital—especially when it was high-tech companies that the two partners planned to start. It was one thing for companies like HP to have large engineering R&D budgets. But Intel aside, restaurant chains were what was hot. The recent bear market on Wall Street and unfavorable capital-gains laws, too, made a venture fund look unattractive. It took Robertson, and the two engineers he arranged in business matrimony, four months to come up with the other $4 million.

Their road shows for potential investors were an adventure—two Willy Lomans roaming the small cities of America. Ever the control freak, Perkins was always at the wheel. He said Kleiner was the world's worst driver, starting to pull off at every highway exit and then coming back onto the road at the last second, over and over. "On my first time with him behind the wheel, I was certain we were going to die," Perkins said. In northern Wisconsin, after getting $1 million from Sentry Insurance, Perkins cold-called Employers Insurance of Wausau. Perkins mentioned to the representative that he'd spent a year in the town as a child. According to Perkins, the rep was so excited that anybody who had left Wausau actually came back, even for a sales call, that Wausau agreed to kick in $1 million to the fund. Another company, on whose board Hillman served, agreed to meet. Recalled Kleiner: "The guy in charge asked us a series of questions like, 'What kind of companies will you invest in?' We said, 'Successful ones.' He kept pressing for specifics. We finally mentioned a word-processing typewriter." The prospective investor thought they were crazy to compete with IBM. He threw Kleiner and Perkins out.

They ultimately got their money—from fewer than a dozen investors ("limited partners," as they were called, since the general partners of the venture-capital fund, Kleiner and Perkins, made all the decisions). The investment stake included $1 million from Rockefeller University, nearly that much from two insurance companies, and the balance from trusts and wealthy individuals (though, interestingly, not from Hewlett or Packard, for Perkins did not want to be seen as their "puppet"). "People who could easily part with their money," as Perkins put it. And always "legitimate sources," which meant money without any taint of organized crime or gambling (not that any venture capitalist took money from those sources). "We were too risky for the mob," Perkins joked. He and Kleiner themselves each put up $100,000. Kleiner & Perkins (KP), as they called both their partnership and their inaugural fund, eventually set up shop in a Menlo Park office complex down the peninsula from San Francisco, and hired a part-time secretary and bookkeeper. Their address was Sand Hill Road, which would attract legions of other venture-capital firms and become the Fort Knox of burgeoning Silicon Valley. Perkins, in particular, became the archetype of the new wealth—and how to go about spending it.

––––––––––

In 1958, the year after he'd arrived in California, Perkins bought his first boat—a modest seventeen-foot sloop made in Hong Kong before World War II named *Pequod* (taken from the name of Captain Ahab's whaling ship in *Moby Dick*). It was a Teak Lady design—a scaled-down classic, varnished, made entirely of teak, with a raised cabin, two berths, and an intricately carved tiller with the head of a dragon and inlaid amber stones. It cost one thousand dollars. Perkins raced it in San Francisco Bay and won a championship his first season. This was much more satisfying than competing in a mere dinghy at MIT. Perkins had the sailing bug for sure.

And because it was made of wood rather than the newfangled fiberglass of the time, *Pequod* allowed him to indulge his need to tinker. Boats, as he was learning, gave him the chance to be alone, to be in charge, to escape, and to compete. A beautiful boat, even if only seventeen feet, was even better: *Pequod* was a hint of things to come. In many ways, boats were his first love. Before he met Gerd and even after—before he was rich and even after—boats provided a refuge, emotionally and otherwise.

Two years later, for five thousand dollars, he went twice as big, buying a thirty-three-foot International One Design wooden sloop from Norway called *Undine*. As with *Pequod*, Perkins didn't rename the boat. Although

a rationalist by instinct and an engineer by training, he believed in nautical superstitions enough that he usually didn't challenge them—and high
on the list of superstitions was that you never changed a boat's name,
since after all, it was a living thing. Rabbits were another part of the folklore that you didn't mess with. Everybody knew that rabbits were bad
luck on a boat. The superstition, like most, had a basis. In the early days
of the British navy, cooks brought rabbits aboard as a food source that
nicely replicated itself. But the rabbits escaped from their cages, made it
into the bottom of the hull, and ate the oakum that caulked the seams
between the planks. The ship sank and everybody drowned, including
the rabbits, who were really bad swimmers. Even Perkins had his doubts
about rabbits. The night before the awful collision at the Saint-Tropez
regatta, the crew of *Mariette* was teasing the chef about rabbits. The chef
was deeply superstitious, believing that even the mention of rabbits could
jeopardize a boat. He implored the crew to stop. Instead, a deckhand
handcrafted some rabbit ears and danced about the deck. The next night,
after the accident, the chef admonished the crew members that they were
to blame: "rabbits yesterday, death today." Don't expect rabbit stew as
your main course on a Perkins-owned boat. The *Falcon* crew was even
forbidden to watch Bugs Bunny on satellite TV.

Perkins raced *Undine* for a decade, finishing near the top of the circuit
in the latter years. Bigger was definitely better—a totem of his business
accomplishments that he delighted in people taking notice of. *Undine*
was more of a project than *Pequod,* and Perkins spent much of his weekends tinkering with the boat at her dock in the Sausalito yacht basin, next
to the Golden Gate Bridge. The boat was a passion. In many ways, it
provided better company than people, since people required small talk
and intimacy, and they were unpredictable besides. Sailboats, too, were
fickle, but they permitted Perkins the semblance of control—and Lord
knows, they didn't ask stupid questions. If there was one thing Perkins
couldn't stand—and there were many such things—it was stupid questions. His Harvard friend, Dave Thompson—who was now living in

the Bay Area—remembered one time at the public dock when Perkins was diving under his boat to clean off the keel in preparation for a big race. An elderly woman wondered what he was doing beneath the boat. "Ma'am," Perkins replied, "I keep my bicycle down there." The poor woman thought he was crazy. For Thompson and so many other friends, it was a perfect example of Perkins not suffering fools.

Even before KP got itself rolling, Perkins cast off *Undine* for his first yacht—the forty-seven-foot wood yawl *Copperhead*. (A yawl was fore-and-aft rigged with two masts, the smaller of which was behind the rudder). With her unusual aft cabin in addition to the main cabin, *Copperhead* was designed by the well-known designer, Philip Rhodes, whose clients included the Rockefellers. Rhodes's *Weatherly* won the 1962 America's Cup. *Copperhead* cost Perkins $50,000 in 1970, which was a lot of money back then (the equivalent of about $260,000 in late 2006). Remarkably, in those days *Copperhead* was one of the biggest boats on San Francisco Bay. The only two bigger boats in the bay were seventy-two feet long—among the largest sailing yachts in the country other than remaining J-boats. "The seventy-two-footers were the biggest things I'd ever seen in my life," Perkins said. "I couldn't imagine owning something like that. It wasn't something that occurred to me." That had been especially so before his University Laboratories home run, though Perkins said any big boat would have been considered anachronistic. The J-boats were born of a time of low income taxes and carefree attitudes. To go big in the early 1970s, Perkins said, would've been bizarre.

He raced *Copperhead* offshore—along the California coast and down to Mexico—but mostly used the boat for cruising with his family around the Bay Area and inland toward Sacramento through the Sacramento River Delta. Though Perkins sold *Copperhead* long ago to one of his venture-capital partners, John Doerr, Perkins felt a strong emotional bond to the boat—his entry into the world of high-end sailing. To this day, he retains a nominal ownership percentage in the yacht. His insurance broker might question the wisdom of his sentiment. Doerr was notorious in

Silicon Valley for his erratic habits behind the wheel of a car—so much so that he wisely chose to hire a full-time chauffeur—which raised the question of why KP partners apparently cannot drive. What's not well known is that he's no better at the helm of a vessel. While *Copperhead* has a professional captain, Doerr on occasion still took the helm—and wound up endangering buoys, bridge pylons and a variety of marine mammals. Perkins must really love *Copperhead*.

FOUR

Prophets and Profits

Kleiner & Perkins had big dreams and an $8 million pot, yet initially could find little to invest in. The days of young engineers rushing the front door with business plans in hand was still far off. "It must be really hard to decide whom to fund among so many choices," Henry Hillman once remarked to Perkins. Perkins couldn't resist a wise-ass response: "It's easy—it's almost all your money!"

For the first two years, KP backed almost nothing in high-tech, instead putting money in the kinds of deals that wind up being ridiculed on the front page of the Wall Street Journal: a combination snowmobile-motorcycle—the "Snow Job"—that came out in 1973 during the first oil embargo, and then a business that would retread sneakers. Those two turkeys alone ate up $650,000. A waste-treatment company blew another $1 million in three years, only to be left, in venture-capital parlance, "at the bottom of the ocean." (VCs seem to love maritime allusions.) Kleiner was from Vienna and adored its pastries, so he proposed backing a can't-miss idea called Mr. Bumbleberry's Pie. Then he tried a piece—this was "field research"; Mr. Bumbleberry didn't get his money. Another troubled company, Dynastor, taught the partnership the perils of luck and timing. Dynastor aimed to revolutionize the market for floppy-disk drives that were used for computer memory. KP put nearly $500,000 in the company, which was never quite fast enough to keep up with competitors. The moral to Kleiner & Perkins was to get out of bad deals sooner. Even though Dynastor returned a profit of a few hundred thousand dollars, it had squandered a lot of time. KP maintained an "ICU (intensive care unit) list" made up of companies in need of emergency treatment. But the ICU didn't stick with a patient forever. When the plug needed to be pulled, the partners prided themselves on not showing any sentiment.

The point might have been obvious, yet the VC landscape was filled with examples of good money chasing bad. Kleiner & Perkins aimed to formalize certain principles of doing VC—raising the money, evaluating proposals, finding talented management. Whereas entrepreneurs were supposed to be fanatics ("drinking the Kool-Aid" was a favorite metaphor), venture capitalists were supposed to be agnostic, though urgently impatient at the same time. They didn't care so much if the widget changed the world—though that was a nice by-product—as long as the widget produced double-digit returns. Better to double your money on mediocrity than lose it all on a dream. An entrepreneur might be dancing away on the *Titanic* even after she hit the iceberg; the VCs would be first in the lifeboats.

What Gene Kleiner and Tom Perkins set out to do was create a dispassionate system for winning more often than losing, even if instinct still ruled their decisions in the end. In the process, they invented a new vocabulary of money, even if the system sometimes looked like a crapshoot to outsiders. Most important, they were aggressive in managing the companies in which they invested—a direct, grease-under-the-fingernails predilection based on their backgrounds as technologists and managers, experience that even a shrewd financial analyst like Arthur Rock didn't have. It was Rock whom the young firm wanted to distinguish itself from. Other VCs turned money over to an entrepreneur and then just watched from the grandstands. This is what wealthy families like the Rockefellers or the Whitneys or the Hillmans had been funding for years. They called their investments "risk capital," before "venture capital" became a part of financial parlance. With Edex and University Laboratories, Kleiner and

Perkins had each demonstrated they could take a venture all the way from idea to start-up to profits to liquidity. When one enterprising physicist came in with an idea for an acoustic-imaging company, he just wanted to talk about rates of return. Perkins wanted to discuss was acoustic diffraction. The entrepreneur thought he had met a madman. They did the deal, and Acuson was born—a pioneer in diagnostic obstetric ultrasounds.

KP was not going to be a sidelines investor. That meant serving as architect, hustler, tactician, manager, and, when necessary, executioner. Anybody could write a check, but KP viewed money as the "least differentiated" product in the world. "We were the masters of meddling," Perkins said. "We were famous about calling the entrepreneurs weekly if not daily. We might even be more ambitious than the entrepreneurs were. We were part of a team—whether they wanted us to be or not." Entrepreneurs often fixated on what percentage of their company they would own—whether they had "control." Perkins sneered at them. "I gave them full disclosure," he said. "I told them that percentages didn't matter—that we controlled their fate regardless. If we were their lead investors—even with, say, 20 percent of the company—and the entrepreneurs need more money, which they would, then if we didn't give it to them, this would be the kiss of death, because nobody else would either, which they had to know because they came to us in the first place." It sounded a bit like the Mafia—partners for life in a deal you couldn't get out of.

That M.O. was the reason KP never got to make a deal with Steve Jobs. After he lost a power struggle at Apple Computer in the mid-1980s—years before his triumphant return and second act at the company—Jobs went off to start NeXT Software. He and Perkins both wanted KP to back it. But Jobs insisted that he alone control KP's ability to "get liquid"—to take the company public or to sell it. KP never allowed an entrepreneur to dictate those fundamental strategic calls. "Trust me," Jobs assured Perkins, who said he did, but that while Jobs's concern was understandable, it was naïve. The VCs always ran the show, according to Perkins. "If you succeed, you'll need money. If you fail, you'll need money. Either way, you'll need money—and you'll have to deal with us. Even if you go public, we may not sell our shares. We may be on your board of directors twenty years later."

Ultimate control is why KP, as well as other venture-capital firms—in return for its deep pockets—demanded a substantial ownership percentage of any company, as well as a seat on the board. It could turn out to be a Faustian bargain. VCs will say their allegiance is to the founders and to the CEO, but they're there primarily to protect their investment. Most times, all three coincide. When they don't, there's bloodshed. Perkins could be an outstanding management assassin. (When the time came in 2005 for the Hewlett-Packard board of directors to get rid of its then-CEO, the controversial Carly Fiorina, it was Perkins who was brought out of retirement back on to the board to deliver the coup de grâce.)

Even after a company went public, KP didn't go away, holding on to its board influence. It was one thing to be a trusted adviser or corporate psychiatrist, but the time often came when the VC had to be the tough guy who fired the entrepreneur-turned-CEO ("redeploying the founders" was the euphemism) or sell out to the highest bidder (thus the tag "vulture capitalist," coined by the normally mild-mannered Gordon Moore of Intel). The "system" of venture capital required knowing when and how to play rough with entrepreneurs—exploiting their inexperience and emotions, pitting founders against each other, and negotiating tough prices and generous ownership stakes that in hindsight infuriated the entrepreneurs. If entrepreneurs attempted to play competing venture firms against one another, then the VCs might band together in a deal. That meant a lower amount of investment per VC firm, but it also kept the price low—not a particularly free market, but pretty smart when they got away with it. Because of KP, venture capitalists as activists—rather than bystanders—became the model in Silicon Valley.

KP's guidelines took on a kind of mythology within the Valley, as venture capital became a multibillion-dollar industry. There was Kleiner's First Law: "When the money's available, take it," which became in other tellings Kleiner's Law of Appetizers: "When they pass the hors d'oeuvres around, take two." Neither Kleiner nor Perkins remembered saying either, but then Yogi Berra doesn't recall saying most of the things attributed to him. And there was Kleiner's Second Law: "There is a time when panic is the correct response," which was another way of saying to get out of bad deals quickly. And Kleiner's Third Law: "Never sell unless there are two buyers." Or another Third Law, which was counterintuitive and brilliant: "If a decision is incredibly difficult, it doesn't matter what you decide." And an unenumerated KP law: "Make sure the dog wants to eat the dog food." If there weren't any customers for your wonderful new technology, you were a failure.

Perkins's best-known law was: "Market risk is inversely proportional to technical risk," which was quite insightful, because it placed value on products that looked impossible to develop. If the product was easy, presumably others would come up with it and crowd the marketplace (as KP-funded Netscape learned from Microsoft in the late 1990s). More difficult products, when they worked, could dominate an emerging industry. As a rule, Kleiner & Perkins never wanted to face both market risk and technical risk in a start-up. It almost never paid to be a follower. An old Valley joke, which Perkins cited as emblematic of the way *not* to do business: "What do you get when you cross a lemming with a sheep?" Answer: a venture capitalist.

As attractive as a risky product was—given that technical risk signified immense market opportunity—Perkins prided himself on identifying and reducing risk at the outset. The former wasn't always so easy: a concept or product didn't come with a warning label. And even if you identified potential flaws correctly, you couldn't eliminate risk altogether— it was inseparable from the practice of venture capital. *Venture* implied unforeseen circumstances. But if Perkins was going to take a chance—on a new product, on a new industry, on some harebrained commercial scheme—he would do all he could to minimize the risk and do so as early as possible. If he was going to lose out, he wanted to "fail fast."

Of all the various types of risk that a new venture faced—technical risk, marketing risk, manufacturing risk, personality risk—it was technical risk that Perkins most preferred. Given his background, this was the kind he believed easiest to overcome and where he hoped to direct any early investment. Once most of the risk was contained, KP poured in larger sums to let a company ramp up. Though more than a few entrepreneurs complained that KP negotiated with brass knuckles—holding out for a larger ownership stake at the lowest possible price—Perkins said price was the least of his worries when structuring a deal. A reputation for *over*-paying even had benefits, the logic went, because it guaranteed KP got to see most deals. Like all venture capitalists, Perkins was a gambler within limits. The only way to get really rich, to build empires—and Perkins surely viewed himself as an empire-builder—was to place bets. But by militantly trying to reduce his risk, be it a hundred thousand dollars or a million, at least he could avert most disasters. It might mean walking away from an investment completely, even when it still somehow seemed to hold promise. But that is what constituted venture-capital discipline in Perkins's mind and much of what he believed set his approach apart. When the time came to imagine and build the *Maltese Falcon*, it was this risk-reducing methodology that Perkins followed.

———

Due largely to its initial inability to find other people's high-tech ideas to bankroll, KP decided to incubate its own. Rather than be reactive—waiting until an entrepreneur walked in with a plan—KP aimed to see where the world was going and how the partners could invest in it. This self-assurance—some called it arrogance—became a hallmark. KP partners liked to say they had a list of five companies in their back pockets that needed to be started; for many of them, a partner might even serve as CEO for the first few years.

In 1973, the firm had hired Jimmy Treybig, a Hewlett-Packard electrical engineer by way of the Texas panhandle and Stanford Business School. He was the first of the young "associates" brought in by KP to bird-dog deals and pursue their own entrepreneurial ideas. Treybig wanted to sell to businesses a foolproof minicomputer that saved data even when it broke down. The trick was to utilize overlapping dual microprocessors so that even if one crashed the other backed it up. Banks and airlines and stock exchanges, for example, were desperate for a reliable "non-stop" fail-safe computer system. Treybig got KP to kick in $1.5 million—a big chunk of the first KP fund—for a 40 percent stake and Perkins as chairman of the board. In three years, when Tandem Computers went public in 1977, the value of the company shot up more than eight times. By 1981, Tandem was worth more than $220 million and was ranked the fastest-growing public company in the country.

Treybig liked to accentuate his Texas heritage by noting that in Houston "you start playing poker at the age of six." Perkins, though, saw him as a rube and took him on as a human-resources project. When they both went to Wall Street to promote the public offering, Perkins decided to take him to Brooks Brothers. He bought Treybig a suit, shirts, socks, and shoes. Treybig was entirely presentable. But at his first meeting, Treybig announced, "Look—Tom dressed me!" Treybig was irrepressible and that style worked wonders. Perkins laughed, and the two of them worked well together for years.

Yet, over his long career, Perkins often had vitriol for those that stood in his way or that he otherwise disliked. At any given moment, that list could include opponents, politicians, academics, journalists, doctors, lawyers, insurance agents, historians, authors, yacht brokers, yacht snobs, seasick guests, princes, kings, actors, housing contractors, real estate agents, craftspeople, restaurateurs, flight attendants, the French, writers of a *Sopranos* episode he didn't care for, and folks who didn't appreciate his tall tales or his elaborate, clever, and sometimes mean practical jokes. Patricia Dunn, the chairman of the board whose ouster Perkins engineered in the fall of 2006 during the Hewlett-Packard boardroom spying scandal, never forgot the time Perkins buzzed her at a dinner party with his prized remote-controlled toy helicopter. She was talking shop with HP's newly crowned CEO, Mark Hurd. Perkins was bored silly. So were the other guests. So Perkins made several passes with his helicopter at Dunn's head.

Perkins's unfavored list encompassed anyone who he thought wasn't as smart as he, which covered a vast spectrum. His disdain may have been inextricable from his competitive core. But part of Perkins's charm was that even as he embraced his mammoth ego, he made fun of it and on occasion tried to tone it down. A KP partner once suggested that, on a trip to the dentist, Perkins consider having his fangs pulled. Yet there was one category of professional for whom he held special contempt, a scorn that began when Tandem went public. Any initial public offering required investment bankers, who underwrite the IPO by finding customers and agreeing to make up any shortfall.

Perkins thought the whole arrangement was ridiculous, since he believed his typical stock offering had sufficient buyers anyway. He never tired of noting that "investment bankers" had nothing to do with investments or banking, and that "fee-grubbing middlemen" was a more apt term. He loved to recount the Thanksgiving story about the Tandem offering. The holiday was nearing and Perkins was to speak at a lunch of investment bankers. One of the senior partners of the banking firm, Tommy Weisel, turned to Perkins and told him it didn't matter what he said. "Just stand up, Tom, and go, 'Gobble, gobble, gobble!'" Apparently, investment bankers would do any deal, as long as they got their fee. That's how little respect the partner had for his own people. Venture capitalists, too, had their eyes on the financial prize, but in the world according to Perkins, the VCs at least thought in terms beyond the next day. Investment bankers, his view went, couldn't think beyond their next meal. Quoting one of his partners, Perkins said they have "all the self-restraint of lobotomized sharks." But, Perkins noted, long after the sharks were gone, "when it's down to just cockroaches and investment bankers, it won't be the cockroaches that are left standing."

His cynicism about investment bankers, and ability to hold a grudge, led to one of them getting excluded from some KP financing deals. And it grew out of a personal dispute that seemed to justify Perkins's attitude. Perkins had established trusts for each of his two children, into which assorted stock from KP venture funds flowed. The trustee on both accounts was Frank Caufield, his longtime partner and close friend. Hundreds of shares kept flowing into the trusts. In the summer of 1989, Caufield told Perkins that nobody was keeping track of all the stock, and he proposed liquidating most of it. Perkins agreed. Caufield called the Morgan Stanley broker on the trust accounts and put in sell orders. Caufield's intention was to sell one thousand shares of each of the many companies, but he mistakenly put in sell orders for ten thousand shares. Three days later, the Morgan Stanley broker called up to inquire, "Where are the shares?" Caufield had accidentally "sold short"—disposing of shares he didn't have, which was a well-known trader's gambit for a quick profit. It worked great, as long as the stock price was going down.

Unfortunately, the prices of the companies were going up. When Caufield was informed of the error, the loss stood at $400,000. He came to Perkins and offered to cover it personally. Perkins would have none of it, suggesting the loss be covered by four parties: himself, Caufield, the children, and the broker. Apart from thinking that Morgan Stanley would happily agree to that compromise, Perkins thought the firm was most culpable, as it should've known better not to execute the transactions, for trusts don't typically engage in short-selling. He also thought it wouldn't be an issue, since he was chairman of Morgan Stanley Ventures, an arm of the firm that provided financing to young companies after they had survived their start-up stage. KP was a big client for Morgan Stanley and the firm's venture arm made it even more money. It was in Morgan Stanley's interest to keep Perkins happy. Caufield liked the plan. Morgan Stanley did not. When he was next in Manhattan, Perkins went to see the chairman of Morgan Stanley. If the firm didn't relent, Perkins threatened to quit Morgan Stanley Ventures immediately and to do everything to see that no KP business came Morgan Stanley's way. "I can be very vindictive," he told the chairman, which was something that anyone who knew Perkins took seriously. Perkins rarely went nuclear, but when he did—as Hewlett-Packard's Patricia Dunn learned—the body count piled up.

"But it was Frank Caufield who made the mistake," the Morgan Stanley chairman countered to Perkins. "If we covered every mistake made by every client, we'd be bankrupt."

Perkins never got a penny back. As he'd promised, he quit Morgan

Stanley Ventures, and he urged his KP partners to send no more business to Morgan Stanley. While some KP deals made their way to Morgan Stanley anyway, Perkins was able to reduce the flow. In keeping its $100,000 that day, he concluded, the investment-banking colossus passed up millions down the road.

————————

After Treybig left KP in 1975, the firm hired another young entrepreneurial gun named Bob Swanson. Like Perkins, he had both a science background and an MBA (both from MIT). Like Perkins, he believed in the power of technology and technical ideas. And like Perkins, he had no hesitation about the goal of creating companies. It was all well and good to want to change the world, but in their philosophy, making a corporate profit was admirable and entirely consistent with bettering mankind.

Swanson had come to California to run Citibank's West Coast venture operations. But the twenty-eight-year-old didn't want to manage other people's money. He wanted a company of his own and approached Kleiner & Perkins, whose incubation of Tandem had achieved notice around the Valley. Perkins suggested he join the firm for a few years to work through his ideas. Swanson settled on the pioneering area of gene-splicing—recombinant DNA technology, as professional journals called it. In Swanson's view, biotechnology was just another form of information technology: the information just happened to be DNA, stored not on chips but on genes. In the mid-1970s, genetic engineering was theoretical and the locus of research was academia. Kleiner and Perkins knew nothing of the nascent science and were dubious of Swanson's ambitions.

Sniffing around for gossip, Swanson heard that existing companies were barely beginning to think about a way to commercialize gene-splicing—moving a piece of DNA from one bacterium to another. With the correct coding sequence, the second microorganism reproduced and passed on the new traits to its offspring. Presto—a genetically altered creature! While the ultimate object might be replicating a human gene—

and cloning an individual—in the near term, if successful, it would be ground-breaking in producing man-made hormones and other biological agents. But mass-producing recombinant DNA simply was not part of the commercial or academic mindset at the time, and this provided Swanson his opening. He asked KP to back him in a biotech start-up, with $1 million in seed money—for a lab, technicians, and equipment. In keeping with its policy of trying to overcome technical hurdles early and as inexpensively as possible, the firm balked at that much money. "Wondering if God would let you make a new life form struck me as rather high-risk," Perkins recalled. "Especially in an $8 million fund where we had already had some very significant losses. It occurred to me that the best way to manage the risk of Swanson's idea was to just tell him no." Even so, Perkins proposed subcontracting out different parts of the experiment to existing companies.

Swanson phoned around to various researchers working with recombinant DNA. One of them was molecular biologist Herbert Boyer of the University of California at San Francisco medical school, an early recombinant-DNA patent holder who'd been dubbed the "genie of genes." It didn't hurt his press clippings that his bushy mustache gave him a resemblance to Einstein. Boyer agreed to meet Swanson for ten minutes on a Friday afternoon. In one of those celebrated, apocryphal world-changing Silicon Valley conversations that actually took place, they wound up talking for three hours over beers at Churchill's Pub—academic and businessman, the ideal complement for an industry they would create. Boyer was skeptical that commercializing the technology was possible anytime soon. But Swanson kept pushing and convinced Boyer that with sufficient private funding, the timetable could be greatly accelerated. With an initial $100,000 from KP—in return for 25 percent ownership—Swanson and Boyer founded Genentech in early 1976. Perkins was chairman, as usual. There were no other investors out of the gate. KP soon enough provided another $100,000 in capital. The field of biotechnology was born. Genentech was the first biotech company. More than a thousand followed over the next thirty years.

Using University Laboratories as the model—he said that experience was the foundation for all his venture work to follow—Perkins prevailed in his suggestion to have Swanson and Boyer farm out discrete parts of the first experiment. The point was to see if Boyer's technology could work—the "proof of principle"—without spending millions to find out. Perkins had no interest in renting a big facility, hiring hundreds of employees, and buying lots of lab equipment before attempting the experiments Boyer proposed. UCSF, Caltech, and the City of Hope Medical Foundation each got a separate research contract. None of them was told about the overall project. If the experiments didn't pan out, the damage would be limited to $200,000. If the experiments worked, then Perkins would put up larger sums of capital in return for additional, though smaller, slices of the company, which in the long run meant that Swanson and Boyer ended up with a larger percentage of Genentech than if KP committed to a big check on Day 1. (The only bad part of the plan was that involving the three outside institutions led to patent lawsuits that dragged on for decades.) Within a year, Genentech had a break-through, manufacturing somatostatin—a human brain protein—and then the more complicated hormone, human insulin, for treating diabetes. Perkins's strategy of containing risk by attempting to make just one or two products had paid off.

———————

Genentech went public in October 1980—the hottest Valley IPO since Intel nine years earlier. It made the front page of the *San Francisco Examiner* in a banner headline, GENENTECH JOLTS WALL STREET. Kleiner & Perkins's original $200,000 stake was worth $160 million—an 800-fold explosion in four years. The partnership never had to put in more capital, for it already had its stake and Genentech had plenty of cash. "We owned such a huge amount of the company for so little that it was embarrassing," Perkins said. He had high school friends he hadn't talked to in decades calling him to get in on the deal. (Alas, none of them were KP investors.) In its first ninety minutes of trading, Genentech (stock symbol: DNA) went from $35 to almost $89. Driving up the stock was the public rather than institutional investors; there were buy orders from around the world. Perkins's secretary at the time pleaded with him to find her some IPO shares, which were in short supply. He found some through the investment bankers and, on her behalf, sold them on the morning of the IPO before he even got into the office. The shares had gone up $15 at that point. By the time they both arrived at work, the shares were up another $30. "I got you out at $50," Perkins told her. "I'm sorry." She made out just fine. Like other KP-brewed companies to come, Genentech perfectly captured the public's fascination with a new technology, despite its absence of profit. Genentech is still a well-known corporate name, probably as much because of the IPO splash as its medical and scientific achievements.

Critics derided Genentech as "a science-fiction stock," but even that put-down imbued the company with a futuristic glow. And Perkins, as well as Kleiner, had moved from being independently wealthy to fabulously rich. Genentech was their most lucrative deal ever—"the perfect deal," Perkins called it more than a generation later. From the date of the public offering until early 2007, Genentech's value increased 550 times over, on top of the 800-fold increase of Day 1. Perkins and Kleiner held on to their stock, its value skyrocketing year after year. "It was the wonderful faucet that never turned off," Perkins said. They were now multi-multimillionaires.

That's how venture capital operated, when it produced home runs. It was—and is—a glorious system that might make a few hedge-fund managers of the twenty-first century envious. Venture capitalists made their money in two ways. First, KP's general partners—at the outset, just Gene Kleiner and Tom Perkins, and even today, no more than a dozen or so individuals—annually received 2 percent of the money that investors had committed to a venture-capital fund. In the first KP fund, that meant 2 percent of only $8.1 million, but subsequent KP funds into the 1990s and beyond ranged close to a billion dollars. Partners were entitled to that 2 percent "management fee" regardless of how the fund was doing—a toll gate, as it were. So, even though the inaugural KP fund went through $1 million for a waste-treatment facility that yielded nothing, the partners still took home $20,000 a year for that blunder alone.

After the fixed percentage came the better part: the partners got a piece of the action. This was the "carry"—the lingo for "carried interest"—which in English meant "our cut" of all the profits. This is where the real money was, and the sky was the limit. The carry for the first KP fund was 20 percent. The investors (or "limited partners") got the other 80 percent; these typically were large institutions such as universities and insurance trusts, as well as wealthy individuals. For a few venture capitalists to divvy up 20 percent was to reap a vast personal windfall. (As an ethical matter, Kleiner and Perkins decided that they could not take their piece until the initial capital of the limited partners had been returned in full. Kleiner and Perkins also decided that the respective holdings of all partners, general and limited, would be treated equally. That way, no limited partner would get a phone call like the one Perkins received from the principal of a cattle tax-shelter that Perkins was in. "Tom, *your* cows died," the principal told him.)

Genentech made Bob Swanson even richer than the venture capitalist who backed him. But the personal relationship didn't fare as well as the financial one. That wasn't unusual in Silicon Valley: VCs and entrepreneurs famously had falling-outs. When the VCs decided it was time for the founders to go—when a start-up had matured into a real business or an industry behemoth, and required new leadership—the founders unsurprisingly chose to differ. In the ensuing struggle for control, the venture capitalists invariably won. As Genentech went through tough times and eventually was acquired by the Swiss drug maker Roche, nobody was looking to force Swanson out. After the deal, Perkins even gave up the Genentech chairmanship to Swanson to make up for other chips Swanson lost in the transaction.

Yet he and Perkins weren't the friends they had been. There were disagreements in the boardroom. And there were petty jealousies over such

things as the fancy house Swanson decided to build, modeled in part on
Perkins's place in Marin County. That annoyed Perkins's wife. It didn't
matter that Perkins had called Swanson the best entrepreneur he'd ever
worked with, or that without Swanson, the venture firm might never have
turned out so well. The Perkins-Swanson relationship thawed after Swanson was diagnosed with brain cancer in 1998 and Perkins tried to guide
him through treatment options—looking for the kind of medical miracle
that Genentech technology had generated. Swanson died the following
year at fifty-two—another in a series of relationships gone sour for Perkins that left him, he said, with a "certain degree of sadness."

Because of the aggressive Tandem and Genentech deals, the first venture
fund put together by Kleiner and Perkins was an astronomical hit. As
arrogant and willful as he could be—about his technical background,
competitiveness, and eye for talent—Perkins was the first to admit that
his real gift was "a sense of timing." The best high-tech idea at the wrong
time produced the same failure as a bad idea in the first place. KP's knack
was getting the big ideas right. "We were the surest thing in the Valley,"
Perkins said. He could buy all the Bugattis and Bentleys he wanted; he
even managed to convince Kleiner to splurge on a Jaguar. Both Perkins
and Kleiner tended to stay out of the limelight, the better to give budding
entrepreneurs the exposure and celebrity many of them craved. Kleiner
was nearly invisible. Perkins was happy to sail his boat; show off his cars;
sit front and center at the San Francisco Ballet, where he was a key trustee;
and do the occasional interview with the press. Yet behind the scenes,
Perkins was the man to see in Silicon Valley—certainly for capital, but
also for strategic advice, for entrée to Wall Street or other power centers,
and as an influence broker. If Perkins blessed a deal—even if his firm
wasn't the main backer—the deal began with an aura of success. If he
actively pooh-poohed it, you were in trouble. On the day that Steve Jobs
was forced out of Apple Computer in 1985, he had lunch with Perkins to

commiserate. Perkins thought Jobs regarded him as somewhat of a father figure. (Jobs, in an interview, expressed great respect for Perkins, but said he doesn't need any paternal guidance.)

In their first six years, the two general partners of Kleiner & Perkins only managed to invest $7 million of the $8.1 million they had raised. Yet ten years after they launched KP, the initial fund was worth a prodigious $400 million—an average annual rate of return of 47 percent, even taking into account the profits paid to KP's partners. That was better than the rule of thumb that venture investors tended to follow: get back ten to twenty times your money within ten years. Of the seventeen deals in KP's fund, seven had lost money. Other possible deals—like a personal computer start-up by Jobs and Steve Wozniak called Apple, which went public two months after Genentech—never happened because the partners weren't interested. "Not a great batting average," according to Perkins, "but the two home runs more than made up for it." The few big winners always erased a bunch of losers and made KP the first brand name in venture capital.

Could it get even sweeter? Thanks to Henry Hillman, it could. In 1980, Perkins had a notion to start a huge $500-million "fund of funds." It wouldn't invest in start-ups, but would merely manage investments in other venture funds—much the way some mutual funds today buy stakes in other mutual funds. Hillman hated the idea and urged Perkins to keep at the venture business and to keep it "small." Hillman volunteered to put $50 million into the next KP pool—and offered to raise the carry to 30 percent. Perkins agreed. In a blink, Hillman was making the KP general partners a lot richer, potentially by many millions. Now, for every dollar KP made for investors, the KP partners got thirty cents, up from twenty cents, which is what every other VC firm in Silicon Valley got. If a new fund of, say, $100 million, returned a profit of $40 million, that meant the partners got $12 million between them. Of course, the partners were still getting their take from the earlier funds. That was a great living— all the more so when the downside to the risky investments was simply getting nothing, since the general partners weren't even the ones putting up most of the capital. In truth, getting "nothing" wasn't possible, since they still got their 2 percent of every dollar invested, regardless of performance.

Venture capital filled a vacuum in the financing universe. It could exist to begin with because traditional lenders like banks considered start-ups too risky. Perkins had gotten a bank loan for University Laboratories based only on his healthy accounts receivable. His initial seed money—before there was a company or customers—came out of his own savings. Institutional investors risked liability to their shareholders if they took on too much risk. It was one thing for Sherman Fairchild and other stewards of Old Money to bankroll a new company—for them, it was as much sport as investment. But college endowments, municipal pension funds, and such—the institutions that controlled much of America's financial assets—couldn't take the chance, even if the typical VC buy-in was only a few million dollars at most. Most state courts had long followed the "prudent man" rule; as a fiduciary matter, the big institutional investors could only put money where a prudent man would—and venture capital wasn't in that category.

The irony was that many so-called safe investments weren't, and many venture-capital deals paid off handsomely. The neat thing about placing a bet on, say, Tandem or Genentech, was that the downside was so much smaller than the upside. You could lose your money only once, but you could make ten or one hundred or one thousand times your investment. It's the absolute reverse of what traditional—and putatively safe—banks do. If its loans perform, a bank makes 5, 10, 15 percent. But just one bad loan can wipe out ten good ones. That's why First Federal Fuddy-Duddy Savings & Trust is so risk-averse—and never gives money to a start-up, which lacks any collateral or earnings. Yet it was the banks that so often seemed to be on every stupid lending bandwagon (real estate, oil and gas, Latin America, Mexico, Southeast Asia). That realization—along with the impressive returns that a lot of early venture capitalists were showing—was largely responsible for eventually easing the restrictions on institutional investors. And that's what allowed venture capital to become part of the Silicon Valley system. More big money came in, which led to more company formation, which led to more successes, which led to more money coming in—a cascade of riches. Perkins and Kleiner were smart enough and lucky enough to tap into this financial transformation at the perfect time.

Perkins never forgot that he wasn't wealthy once upon a time. The boy who dined on salmon loaf and scraped his way through MIT couldn't let go of those memories. He always remembered that when he arrived at HP, Dave Packard and Bill Hewlett invited him to their ranch near San Jose. It was a rite of passage into management to herd sheep for the weekend—and to play poker. While Perkins was flattered to be asked, he neglected to inquire about the stakes—and wound up losing a week's salary, which he could barely afford. Much later, as a trustee of the San Francisco Ballet, he had to go to his share of dinner parties, but he said he detested being around "fourth-generation San Franciscan men who had never worked a day in their lives."

His parents lived long enough to see him score financial triumphs they never could have imagined. Perkins said they were proud of him, yet deeply resentful—prisoners of their Depression-era misfortune. His mother demanded "the first million bucks" he made. He declined, though he gave her what he called an annual "salary." When she died, Perkins discovered she'd wound up saving a total of $1 million of her son's money anyway—squirreled away in savings-and-loan institutions across southern California, where she then lived. His father chafed more. Perkins's mother had long considered him a failure; the better her son did, the more she reminded her husband of it. One of the last times he visited his father before he died, Perkins had just been named to the board of directors of Corning Glass.

"Do they pay you for that?" his father asked over dinner.

"Yes," Perkins replied, "I think it's seventy thousand dollars to seventy-five thousand dollars a year."

"How many days a week do you have to work?"

"See, Dad, it's just a few hours—and just five times a year."

Perkins overheard his parents talking that night in their bedroom. His father kept saying, "I can't stand it. I never earned seventy-five thousand dollars a year in my life—and he's making it for about twenty hours of work. I can't stand it!" You couldn't tell if Perkins recounted the story with empathy or glee or a bit of both.

———

KP's initial successes attracted enough business that it was clear the firm needed to grow. Caufield and Brook Byers came aboard and in 1978 the firm changed its name to Kleiner Perkins Caufield & Byers. That's the name it goes by today, despite the addition in 1980 of John Doerr, who in the dot-com boom of the late 1990s earned for KP partners its most monumental profits. Kleiner retired in 1980 and Perkins became less active in the partnership in the early 1990s when his wife Gerd became ill with the lymphoma that eventually killed her. But Kleiner and Perkins still took home many, many millions. When Tandem finally was sold to Compaq in 1997, it went for $3 billion. Perkins alone made $70 million in the deal. (Compaq was then merged into HP in 2002, which is how Perkins made it on to the HP board and got even richer.)

The money got easier as the venture-capital business matured. Billions could be raised from investors in a matter of weeks and many billions more could be earned if deals were hit out of the ballpark. Internet and biotech entrepreneurs could always find funding for the right idea, but proposals in such trendy areas as "green technologies" and "pandemics and biodefense," became attractive, too. Entrepreneurs came to expect that start-up dollars would always be there and success came to be expected by the limited partners. The parade of numbers was simply astonishing—not that the general public knew anything of it. Unlike most investment vehicles, the private funds that venture capitalists run— no more than a thousand nationwide—are exempt from almost all SEC reporting requirements. In the end, they amount to a mysterious force in the financial kingdom—the invisible hand of capitalism ruling the Valley and beyond—accountable only to the laws of economics. The numbers that emerge in a book like this, or in an investigative piece of journalism, come only from limited partners who are willing to share them, though only on condition of anonymity, lest they be booted from the exclusive group.

As of the middle of 2006, the fourteen KP investment funds going back to 1972 had given birth to 475 companies that employed 275,000 people. Those that were still in existence had a combined market capitalization of $375 billion. Of the 475 companies, 167 had gone public; another 166 had been merged into or acquired by other companies— which is how KP got to cash out its investment. The dozens of limited partners—the universities, insurance companies, and such—had received $17 billion between them. The most remarkable statistic: for the investments during the dot-com boom of the 1990s, KP funds were returning profits that more than doubled each *year*. In the final three years of the boom, the annual return came to 346 percent (even as some man-on-the-street day traders, late to the party, lost out). Over its thirty-five-year history, Kleiner Perkins had produced returns of roughly 40 percent a year, compounded, every year. No wonder that Perkins and other KP partners said their little shop was "the most successful financial institution in the history of the world, unless data from the Rothschilds became available."

Despite it all, Perkins delighted in offering a warning, a tale about a kind of risk he'd never anticipated. This was his "if you think venture capital is a piece of cake" story for the cocktail-party circuit only, since he hardly wanted to advertise that he hadn't foreseen everything.

In the spring of 1999, he had gotten a call from a close friend, Roel Pieper, who was a Dutch entrepreneur and the president of Tandem Computers when Perkins was chairman. "I have the most incredible invention," Pieper told Perkins. "You're not going to believe it." The invention was a kind of "smart card"—looking like a credit card, but with an embedded microprocessor—that would transform the storage of digital data. The card's secret coding system would render obsolete all hard drives, CD-ROMs and other storage media. "The typical smart card has only sixty-five kilobytes," Pieper went on, describing a minuscule amout of storage, barely enough for a few word-processing documents. "What would you think if I told you that I have someone who can put two-and-a-half hours of video, with sound, on a smart card?" That would mean the data was being compressed by a factor of a million to one.

"I'd think you were crazy—that you were violating the laws of physics," Perkins said. But he respected Pieper, who had a degree in engineering and wasn't viewed as a crank. Perkins flew to Amsterdam the next day to see the device and its inventor.

The inventor was Jan Sloot, a TV repairman who had spent fifteen years on his supposed technological breakthrough. When Perkins met him, Perkins thought he looked "vaguely like Adolf Hitler." Sloot was a caricature of a wispy, twitchy, lonely man with visions of grandeur. Perkins was highly skeptical and demanded a demonstration. Pieper had Perkins, Sloot, and others over to his house. Sloot set up two boxes—one for "recording" and one for "playback." The whole thing looked ordinary by any measure of solid-state electronics. There were no moving parts, antennae, or flash-memory devices.

Perkins suggested taping a program from live television, which Sloot couldn't have prerecorded. Sloot readily agreed. He put his smart card in the recording box, connected it to the TV and recorded twenty-five minutes of a Dutch cooking show. He then put the card in the playback box and, as Perkins remembered it, "lo and behold, the show appeared."

"We'll do it," Perkins declared. "I'll finance it right now." Sloot and Perkins signed a contract, and shared champagne and cake. The new company would be called The Fifth Force: there were the four forces of nature, and there was Sloot's device.

The next day, Perkins flew home and was so excited that he couldn't sleep that night. This was an invention that could reap billions. He called Pieper to share his excitement. "I'm going to talk to Rupert Murdoch and others. We'll have video phones and we'll be able to get rid of fiber optics."

"Tom," Pieper interrupted, "he died."

"This is so fantastic, we'll be able to . . ."

"Tom, *he died.* He ate the cake and soon after you left he died. We took him to the hospital, but he was just dead." The evident cause was a heart attack.

"But," Perkins said, briefly pausing, taking in the new information, "we're going to do it anyway, right?"

"Well, sure, but the problem is that the key to this is a compiler that somehow translates the video."

By most accounts, Sloot was slightly paranoid. He had apparently hidden his notes, as well as any software he'd used. Sloot's family desperately tried to help, but could find nothing that made the smart card work. Perkins and Pieper hired detectives, tore out wall panels at Sloot's house, and dug up his garden. It was all for naught. The legend of Jan Sloot lives on in some circles of high-tech—did he really have a revolutionary invention and, if so, who did him in? It remains Perkins's favorite parable about unforeseen perils.

———

By the time he retired as a KP general partner in the early 1990s—he was barely sixty—Perkins was well on his way to amassing a fortune of half a billion dollars. At one point, he was serving on thirteen corporate boards of directors at the same time, as well as on the board of the San Francisco Ballet and other charities. But as fabulous as the deals of the 1970s and 1980s were—and as indispensable as they were to establishing KP's central place in the financial food chain—it was the KP venture funds of the 1990s that catapulted Perkins into the range of the superrich. These funds of the incipient Internet era dwarfed the first KP partnerships. When KP-backed start-ups like Netscape and Google cashed in on the public's feeding frenzy, Perkins made off like a Barbary pirate. Netscape was worth $2.3 billion on the day it went public in 1995, launching a contemporary rush to El Dorado. KP's seed money of $5 million instantly became worth

100 times that. KP's stake in Google, $12.5 million, was worth 150 times that on the day of the IPO.

By early 2007, Google was the third-most valuable tech company in the world, behind only Microsoft and Cisco. Had Perkins stayed on as the managing KP general partner, he'd have made even more money. Though he had no regrets—he had pulled back to care for Gerd—he acknowledged the occasional wince at what might've been "whenever I had to come up with another $25 million to plug into the *Falcon*." He wasn't complaining. Google was plenty good to him. Perkins said he could have called his boat the *Maltese Google*.

Google actually represented an end of the gravy train for KP. Without Google, the late-1990s KP investment fund that backed so many dot-com bombs—the $460-million KPCB IX—would have been the firm's first loser ever. Before Google went public, KPCB IX's rate of return was more than 20 percent in the red. "Google was the peach in a bucket of piss," Perkins said. It gave that fund an overall four-to-one gain—pretty good by the standard of passbook-savings accounts—but still nothing like KP funds historically, which had converted millions to billions.

As ludicrous as such amounts of lucre seemed, it made more intuitive sense than the money being made by CEOs and other executives during go-go times. It was the venture capitalists who backed something that for the moment was worth nothing. Firms like Kleiner Perkins placed a bet that they could incubate or launch a new company from the ether, whereas CEOs and MBA-types only came on later to manage or live off what already was in existence. (Similarly, investment bankers, leveraged-buyout specialists, and hedge-fund traders generated nothing new.) It seemed reasonable, then, that if fortunes were to be made, the initial risk takers should hit the jackpot. And when it was young entrepreneurs who came up with an idea on their own and operated without any venture capital for a while, these founders wound up becoming far richer than the venture capitalists. The boys of Yahoo and lads of Google made off with billions more than the VCs who backed them.

The KP deals of the 1990s weren't Perkins's—they belonged to Doerr, who became KP's first billionaire with them. And Perkins was ambivalent about their success, regardless of all the money that the deals made. Some of that was just understandable rivalry: here was Doerr, the former apprentice, seemingly eclipsing the master. Perkins labeled Doerr a marketing genius "who could convince you the world is flat"—which was both praise and damnation in Perkins's mind. When KP in the late 1990s invested a reported $38 million in the Segway Human Transporter—a two-wheeled battery-powered gyroscopically balanced scooter that George Jetson would've loved—Perkins scoffed. The $38 million was the largest single investment in KP's history—three times its bet on Google— and while the Segway was a fine gizmo, in the world of venture capital it was barely a bunt single, let alone a home run. Perkins liked to say, "When you walk out onto the stage and sit at the piano, you'd better be able to play the piece." He wasn't so sure Doerr always knew how to play the piece—or even where the piano was.

But in his mild critique of Doerr, Perkins was more harshly lamenting that so much of venture investing had turned away from engineering and technical brilliance—the skills that led to University Laboratories, Genentech, and Tandem. And he wasn't wild about how fashionable venture capital had become. "My Kleiner partners aren't even limousine liberals or even Learjet liberals—they've become Gulfstream 5 liberals," he said, noting that he still flew commercial (and leaned Republican). "As they've upgraded their aircraft, they've upgraded their political concerns. They're beyond mere electoral politics, and their political concerns are now things like avian flu and green energy." Those were the areas in which much KP money was going in 2006 and early 2007, along with ongoing investment in the Internet. On its own merits, Perkins found the dot-com mania absurd, though he'd been through other financial bubbles in Silicon Valley—and even though he admitted he "loved" bubbles. "We try to produce bubbles," he said. "Bubbles are great. I've been through them all: laser, minicomputer, biotech, the early Internet. I look forward to more bubbles. Fear and greed are inevitable."

Just before the dot-com bubble burst in 2000, he gave a speech in New Zealand—before a sailing audience at the America's Cup festivities—mocking the waves of investment in companies that made no money and had little business plan other than a Web site and the idea that anything from pet food to perfume could be sold through it. Rolling up his sleeve and playfully looking at his watch, Perkins offered that he didn't know how long the stock market could continue straight upward. "But I think we'll at least get through my speech without the bottom falling out," he joked, before painting the bleakest, darkest picture imaginable. The financial press got wind of the speech and wrote about it. Doerr and Byers were on the phone the next day. "What are you fucking doing?" Doerr asked, as Perkins remembered it. "You're machine-gunning our entire portfolio! When you get back, you're going to come in and we're going to talk about your behavior." Perkins didn't return for a month. By then, as he put it, "the bottom of the paper bag was already wet." He never went in for that talk with Doerr and Byers. The dot-com bust soon followed.

KP made out fine. It had gotten into deals early enough that it was able to cash out before the market plummeted. And Perkins was happy to continue to take his KP cut, which was still ample despite him no longer being a full partner. He even partially benefited from the crash, by anticipating it and selling short in stocks like Cisco that he didn't own at the moment but was due to get in stock distributions; short-selling is a bet against a stock, just the opposite of what you usually do in the market. It can be an extremely risky play. For the most a stock can fall is 100 percent. But a stock can go up without limit and if you short a stock and are wrong, your losses can be limitless. Perkins was that certain the crash was coming. "Vulture capitalist" indeed.

With all that money, and without a full-time job, he would now do the only thing he could imagine. He had done fast cars; he owned real estate in Napa Valley and Lake Tahoe and Britain; he had so much artwork that he had to warehouse some of it. Now it was time to build the ultimate sailing yacht. His appetites had grown with his ability to sate them. Gerd reminded him often that he had the "Big Man complex." "You have too much money," she'd kid him, shaking her head with mock horror.

Actually, she didn't know the details. One time, she opened a statement from the bank, shocked to see the family's financial state of affairs. "We owe the bank $500,000!"

"No problem," Perkins responded. "Don't worry about it." As it turned out, she had read the number wrong. Perkins owed the bank $5 *million* in that transaction, though he didn't mention that.

At any given moment, Perkins didn't know his net worth—and probably wouldn't tell anybody anyway, notwithstanding the pleasure it might bring him to know it was more than they had. But he did know that his wealth was vast enough that it wasn't going to hold back his magnificent yachting ambitions.

FIVE

Hull Envy

In the mid-1980s, given his wealth and his self-professed need for a new project, Perkins grew tired of sailing a humble forty-seven-foot yawl around San Fransisco Bay. He was easily bored by what he'd done yesterday—and might do today or this weekend. In the classic Silicon Valley mindset—born of both California dreams and high-tech swagger—he cared pretty much only about tomorrow. Gerd had soured on California sailing, having taken one too many miserable trips down the cold, foggy coast. She would be spending her time restoring a moated manor house, fifty miles south of London, that they'd just bought from guitarist Jimmy Page, founder of Led Zeppelin. Listed in the Domesday Book and once owned by Henry VIII, Plumpton Place in East Sussex would be the Perkins's haven outside the United States. Its purchase came with a story, of course.

Some years earlier, in the aftermath of one of Page's wild parties, someone was found dead in the moat. Perkins didn't know about it at the time he purchased Plumpton's eighty acres. When Gerd began mapping out her renovations, she couldn't get any workers to go near the house because they said it was haunted. The local vicar suggested an exorcism. Perkins laughed and told Gerd this was really about the local church needing a new roof. But a few weeks later, Perkins flew over from the

United States to observe the rite. Saturday arrived, though the vicar did not. Instead, the archbishop drove up in his VW. "Son," the archbishop told Perkins, "you must be wondering why I am here instead of Father Lawton. It's because the forces of evil may be so powerful that when they're suddenly released, should there be an emergency, I should be present." The day was dark and gloomy—and the archbishop's explanation sent a little shudder down Perkins's spine. Under other circumstances, the archbishop and Perkins might have had a lot to talk about: the archbishop had a PhD in physics from Oxford. For an hour, the archbishop performed the exorcism: incantations, holy water, the works. It even came with a warranty. "Should you encounter any manifestations," the archbishop said, "here's my number." There weren't any problems, but six months later, Perkins got a letter from the vicar reminding him that Perkins was still technically lord of the manor in that area. As such, Perkins was responsible for the local parish—which just happened to need not a new roof, but a new foundation. Perkins provided it.

He still owns Plumpton and spends time there, but it really was Gerd's love. She spent several years creating a twenty-acre woodland garden that now has landmark status (and where her ashes were scattered); she also was devoted to growing roses, and produced a garden at Plumpton known throughout the country. Perkins's passion was yachts and he set out to build a great one—at her suggestion. She was as much a sailor as he was, having grown up in Norway racing small sloops and, as a teenager, winning a national Snipe championship with her brother. "You know," Gerd said to her husband, "you guys have a lot of money. We're going to be spending a lot of time in Europe. Why don't we get a big boat to keep there and entertain friends?" Perkins latched on to the idea in "a nanosecond."

He and Gerd looked at various existing boats and considered several of the great shipyards of Europe—in Germany, England, the Netherlands, and elsewhere. These were first-class builders that were just beginning to get in on the boom in what was then considered a superyacht—sailboats and motorboats over one hundred feet in length. Any of the builders would have been safe, sensible choices for Perkins to make his first big boat. He almost went with Royal Huisman in Holland, the benchmark in quality. Italy didn't even seem like a possibility, its workers not then known for fine craftsmanship in yacht-building. (In the United States, there wasn't any yard remotely up to the task, either in scale or skill. Culturally, it also might have been considered too self-indulgent, which wasn't viewed as a character flaw in Europe.) Then Perkins met Fabio Perini, a new yacht builder from Tuscany. The surprising recommendation came from a close Perkins friend who was the most traditional sailing source of all—Bob Stone, the commodore of the fusty New York Yacht Club. Stone, whose establishment credentials also included being a member of the Harvard Corporation for twenty-seven years, had seen a Perini sailboat and excitedly called Perkins to talk about it.

Nine years younger than Perkins, with dark wavy hair, a deep tan, and a Kentucky cigar constantly in his mouth, Perini had similar dash and charisma, albeit on the quieter side. More important, he had an entrepreneurial streak that Perkins identified with. Many other Europeans who built boats had been doing so for generations. They clung to the past as an article of faith. Yachts were noble and true because they resisted the new. Perini embraced innovation for its own sake: the winds of change swept away obsolescence in other industries—why not in sailing? Perkins and Perini made a perfect pairing—*simpatico*, in Perini's words. Though their transactions involved millions of dollars, they sometimes did them on a handshake. Perkins even learned Italian to be able to converse more effectively with him. Their relationship was such that after the *Mariette* collision at the Nioulargue regatta off Saint-Tropez in 1995, Perini drove the seven hours from Tuscany to show up unannounced at his friend's side.

In the port town of Viareggio, a resort along the Italian Riviera, Perini had been quietly showing off since he was a teenager. His father was in the paper business, as Perinis had been for generations in this region. The family company was Alfonso Perini & Sons. Without formal training, Fabio Perini worked there early on, though he spent a lot of time just fiddling with machines on his own, regarding them as unwieldy and slow. Italy's paper-making industry was revered, but it was antiquated. "I was the difficult son," he said (through a translator). "I always imagined my father let me do what I did because he didn't want to deal with it. I later discovered it wasn't like that. He wanted me to gain faith in my own convictions."

At seventeen, the mechanically precocious Perini invented a method for automatically feeding paper, under tension, into his father's paper machines that converted huge spindles of tissue into individually sized rolls. That feeding task had always been done manually before. Within months, Perini had registered the first of many patents and his father set him up in his own company with fresh capital. At twenty-five, Perini fashioned a high-speed device that easily converted paper into tissue—for toilet rolls, as well as kitchen and industrial toweling. There were two Italian competitors at the time. Perini's company made the manufacturing equipment better and cheaper, and within two years, the rival businesses were gone. He had an effective monopoly in the country and was on his way to becoming a multimillionaire. The two major international competitors—one American and one German—eventually retreated from the market. So by the time he was forty, Perini was the world's dominant wholesale supplier of machines to make tissue products; for example, the Procter & Gamble mills that make Bounty and Charmin came from Perini. With annual sales that reached $200 million, his manufacturing empire extended from Green Bay, Wisconsin to Brazil, Germany, and Japan.

It's hard to develop a passion for toilet paper, notwithstanding the profits (and the jokes about being "flush with cash"). In his spare time—much like Perkins—Perini had discovered sailing after his early success. In 1970, he acquired a seventy-two-foot ketch that he used to explore the Tyrrhenian Sea off the western coast of Italy. (A ketch was just like a yawl—fore-and-aft rigged with two masts, except the smaller mast was in front of the rudder.) From that boat he graduated to a similarly sized schooner that he sailed on trans-Atlantic runs to South America and elsewhere. Perini had previously mastered flying in gliders, grew tired of it, and figured his skills would translate well to the water. It wasn't that easy, he learned.

Sailboats were complicated: sail rigs, particularly on large vessels with multiple masts, required a sizeable and muscular professional crew to manage. The maze of lines and sheets and halyards took constant attention and could be dangerous for anyone on deck. During a tack, a genoa's sheet—the piece of rope that extended from the sail back to the cockpit, where it could be used by crew to control the sail—came whipping around. If it hit somebody, it might injure them or knock them overboard. As much as he liked the sailing aesthetic, Perini found the technology frustrating and dated. "I wanted a boat that didn't need a professional skipper," he said. It would be "a boat whose sail plan I could control myself, on which I could get rid of the winches on deck and lines all over the place and the muscle power to handle it." No one had imagined quite such a thing—a large boat that was easy to sail. So he decided to construct it himself—no matter that he wasn't a naval architect.

The result of that indulgence was the 128-foot *Felicità*—for "happiness"—at the time, 1984, one of the most automated sailing yachts ever, as well as one of the biggest and most luxurious. Using the facilities of a shipyard near his paper-machinery plant, Perini built the ketch with his own funds. His breakthrough was an automated "captive" winch. On smaller yachts, the sheets that controlled the foresails had always been wrapped by hand around circular drums, which could be hand-cranked to assist in sail trim or tacking. These metal winches typically were located near the cockpit; the sheets ran all the way from the corner of the foresail back to and around the winches, and then extended to crew handling the sail. It was a perfectly workable arrangement on a reasonably sized sailboat. But on a large vessel, the sheets were too thick and heavy to maneuver manually, particularly through an obstacle course of gear on deck. And crew members who put their hands near the lines were at risk of having them injured in the event of a mistake. If the line was released too quickly, its speed and heat were sufficient to sever a finger.

Based on his experience with automating a paper machine—which, after all, involved drums and rollers—Perini designed a motorized winch that could handle the lines of a large sailboat—pulling them in or up, and letting them down or out. The trick was a "tensioning arm"—similar to what he had designed for feeding paper machines—that kept a sail's lines taught. Without tension, a sheet connected to a sail jammed or knotted up. Perini insisted on his winch being electric rather than hydraulic, which went against conventional wisdom favoring speed and lightness, but Perini thought electricity would make the design simpler. And he came up with a way to put the winches safely under the deck, controlled not by manual power but by the push of a button in the cockpit. When Perini first took the helm of *Felicità* in 1984, and operated the series of buttons that hoisted the sails, adjusted them on a tack, and then furled them away, it was the dawn of modern sailing. The captive winches enabled the creation of the first sailing "superyachts": without the winch, there could be no realistic way to sail such yachts.

Felicità incorporated other new features. A retractable centerboard within the keel allowed a boat to enter shallower channels; that was an issue because a large sailboat needed a deep keel to maintain stability and sail close to the wind. While *Felicità* was sumptuous, she also was relatively light, for Perini wanted to maximize performance. And vitally important was her "flying bridge"—an enclosed area atop the cabin, from which the boat could be controlled. Such a bridge was common on motorboats, but heretical for a sailboat. It just wasn't what you did: the additional superstructure made a sailboat appear top-heavy, altering its fine lines. Some boats of the J-class era had almost entirely flush decks, with no superstructure at all.

The flying bridge, along with the winch system, became a hallmark of Perini boats—and every Perini had one until the *Maltese Falcon*. It was both lauded and lampooned. Those who loved the flying bridge said it brought convenience to sailing—providing increased visibility and a self-contained area with the wheel, engine controls, electronic buttons, and navigation screens required to cruise or race. Competitors acknowledged that the luxurious Perinis seemed to attract buyers who had spouses who came along for the ride. "A house always sells well to the wife if the kitchen and baths live up to expectations," said Michael Koppstein of the Royal Huisman Shipyard, who was captain of the first Huisman superyacht, the sloop *Whirlwind XII*. The cognoscenti who hated Perini's innovations said it was too reminiscent of a motorboat: if you really wanted an ugly, smelly, noisy stinkpot, just buy a stinkpot. Perini thought that view was foolish—he had as much scorn for motorboats as any blueblood. Perini wryly explained it like this: "I always say that if you encounter a really rough sea in a sailing yacht, you regret having left port. But if you encounter a really rough sea in a motor yacht, you regret having been born." Powerboats were not as stable—without a keel for stability and sails for steadiness, they bounced around in a storm like a bobbing cork.

Perini remembered the naysayers of "easy sailing." They complained even at the invention of something so minor as a motorized self-furling jib (which eliminated the need to detach the sail from the forestay after every use). "Purists objected at the beginning, but they were sailing on boats half the size," he recalled. "I respected what they had to say, but without the new systems, large yachts were impractical." Apart from the mechanical aspects, if these superyachts needed too large a crew, then there would be insufficient room for lavish staterooms and amenities for the owner and guests. Perini confronted the dilemma between aesthetics and tradition on the one side and function and utility on the other—and he concluded that the highbrow yachting community would accept the latter, especially if he was right that bigger was always better. "On a motorboat, you turn on the engines," he said. "On a Perini, you turn on the sails."

———

There was no precise definition of a superyacht. The online publication *Slate* once quipped it was "the classic indicator of someone who has so

much money that he doesn't need to make any more." Some in the yachting subculture would tell you that identifying a superyacht was akin to
Supreme Court Justice Potter Stewart's observation about hard-core por-
nography: "You know it when you see it." Still, in the mid-1980s, when
the term started appearing, most agreed that a superyacht was any very
expensive, privately owned, custom-built, professionally crewed luxury
vessel that was longer than one hundred feet (or roughly thirty meters,
the measure used by most shipbuilders). It came with all the trimmings:
air-conditioning, wine cellar, bar, leather furniture, TVs and other entertainment gadgets, brass and stainless-steel fixtures, hot tub, as well as
staterooms for the owner and several guests. There were fewer than three
hundred such yachts in the world when *Felicità* was launched—and virtually all were powerboats.

When *Felicità* broke through this magic number, it was considered
big news in yachting circles. These humongous yachts reflected advances
in materials and such technologies as Perini's automated winch. Yet they
were more significant as markers of a new age of wealth. Over time—
whether from Reagan-era "greed is good" capitalism, or post-Gorbachev
privatization in Russia, or the dot-com proliferation of the nouveau
riche—a big yacht meant that you had arrived in style, a symbol of success that distinguished the filthy rich from the merely wealthy. Fancy cars,
jumbo diamonds, a mansion so large it deserved its own gift shop—all
these were fine, but a yacht conferred respectability and class. And better
yet, you could take it with you to St. Barts, Cannes, and Porto Cervo, or
wherever else you wanted to be seen. And size mattered—as it always had.
When it quickly became insufficient to be merely big, then yours had to
be bigger. For how shall it profit a man to have a big yacht if somebody
else has a bigger one?

It didn't take long before 100 feet meant little, even to sailors. Aiming
toward 150 feet was entirely predictable. Based on the fact that *Felicità* at
128 feet worked well, and given that the paper business was running at
full speed (he sold the business to a German multinational corporation in
the early 1990s), Perini decided to start his own shipyard in Viareggio.

He called it Perini Navi and he intended nothing less than to challenge established European builders of large, blue-water sailing yachts. Not only would he build the boats—hull, superstructure, masts, and rig—his company would be integrally involved in designing them. "We didn't think of ourselves as a shipyard," said Giancarlo Ragnetti, the CEO of Perini Navi. "We were a design company that went into business only after we had built our first boat. We created a product, and only then a market."

What Perini needed now was a customer. At just the right time, Tom Perkins walked through the door. His previous boat was the forty-seven-foot *Copperhead.* Though she had been much used around the Bay Area, she hardly befit a new Silicon Valley tycoon, whose successes with Tandem Computers and Genentech put him in an elite financial league. Perkins and his wife—she heartily approved of a Perini yacht's capacity for entertaining—agreed to have Perini build a ketch three times the size of *Copperhead,* and at 140 times the price. Perkins welcomed Perini's instincts to push technology to the edge, admiring his technical background. And Perkins welcomed the chance to show up the older shipyards. His $7 million *Andromeda la Dea* would be among the largest privately owned sailing yachts in the world. Compared to most sailboats prior to her, she was leviathan—a scale barely contemplated, let alone built. It would've cost him more, except for a clerical error. Perini's office had accidentally sent Perkins a contract with the price tag of the first Perini boat. Perkins obviously caught the mistake. Perini said he'd honor it, Perkins refused, and they agreed to split the difference, still saving Perkins $1.5 million. The way in which both men handled the situation further enhanced their relationship.

Launched in time for Christmas of 1987, *Andromeda* had three distinct levels and was 142 feet long—almost 50 percent longer than Christopher Columbus's *Niña,* the *Pinta,* or the *Santa Maria. Andromeda's* mainmast was as tall as she was long; her mizzenmast was more than ten stories high. Like most Perinis to come, her hull was painted dark blue with a white superstructure and big windows—not small, round portholes—surrounding the main level where entertaining and dining took place. In technological sophistication and automation capability, *Andromeda* represented a major leap. Perkins knew he'd never actually sail her single-handedly—the solo racing records held little appeal to him and, besides, who would do the gourmet cooking?—yet he liked the idea just the same.

Andromeda's gadgetry was akin to *Felicità*'s, but it further advanced Perini Navi's sail-control technology. For an automated rig to work, there had to be sophisticated software that took into account what the sails and winches and other moving parts were doing and then synchronized them. In heavy air, the sheets behaved differently than in a light breeze. If a line jammed because of a malfunction or an obstruction, that, too, had to be factored in. Perini Navi software monitored ongoing performance and stored the data in onboard computers so that designers and engineers could assess it when a boat came in for regular maintenance. With *Andromeda*, Perkins—now in his fifties, and with a cast of younger partners running the venture-capital firm back home—could begin spending many weeks of the year on board, cruising the Mediterranean in summer and the Caribbean in winter. Along with Plumpton Place outside London, the boat gave him international visibility. It was dandy, while it lasted. But he naturally wanted a better boat—in his mind, there was always another ketch. During a cruise with Fabio Perini and his wife, Perkins saw Perini sketching out a new design. Perkins offered a few suggestions: make it longer, sleeker, faster, he advised. By the end of the cruise, Perkins had agreed to buy the new model. It was no accident that Perini had brought along his sketchpad. Perini was a brilliant designer, and he wasn't bad at marketing either.

Three years after *Andromeda* was built, Perkins took delivery of version 2.0. She was twelve feet longer—crossing the psychological barrier of 150 feet—a foot wider, and carried 11 percent more sail area (including an unusually rigged inverted staysail between the two mainsails that Perkins designed himself). This second *Andromeda* had four big staterooms, double tenders, larger diesel engines, better crew accommodations, and a larger flying bridge—and it came at a price. The second *Andromeda* cost $12 million—more than Perkins had spent on his four prior sailboats combined. Perkins considered her his dream boat. For more than a decade—through the rise and fall of the dot-coms, through the illness and death of Gerd, through his marriage and divorce with Danielle Steel—Perkins kept driving *Andromeda la Dea* across the seas. During a four-year period beginning in 1990, covering more than fifty thousand miles, the boat became the first Perini to circumnavigate the globe: from the Mediterranean to New England to Cape Horn to San Francisco to the South Pacific to Australia to the Suez Canal and back to the Med. (The long time frame allowed Perkins to space out the trip and to spend time in California and England, as well as to continue working.) *Andromeda's* cruising made her Perini Navi's ambassador to the world. With a fire roaring in the hearth of the main saloon, with her hull and rig illuminated at night, she was stunning.

Well after he had sold her, *Andromeda* still reminded Perkins of times good and bad. Under different circumstances, *Andromeda* might have been the last superyacht he owned. When the boat first spent time in the Bay Area, Perkins and his wife entertained aboard her relentlessly. At one dinner party at her berth in Sausalito, they invited Gene Kleiner and his wife Rose up from their home near Palo Alto. The Kleiners weren't sailors or bon vivants and knew little of the yachting scene, let alone a vessel such as *Andromeda*. Kleiner had long ago retired from the venture business and, while he and Perkins stayed in touch, Perkins's obsession with boats wasn't something they discussed. Kleiner lived in a simpler world, where objects of finery extended only as far as his comfortable four-door sedan. As he and Rose walked along the Sausalito dock, they were in a panic when they saw Perkins waiting for them by the starboard ramp of *Andromeda*. "Tom," Kleiner asked excitedly, "how can you afford this?"

"Gene, it's a partnership! I know how much money you have—it's about the same as me. *You* can afford this!" Kleiner had never bought a boat—he remained mystified the entire evening.

Sometimes the adventures of *Andromeda* were grand. And sometimes funny—like the time Patrick O'Brian was a guest. O'Brian, who had written many bestselling maritime novels of the Napoleonic era, had long been Perkins's favorite author. So several years before O'Brian died in 2000, Perkins tracked down his address and sent a letter. Wanting to repay "years of reading pleasure," Perkins offered O'Brian two weeks on *Andromeda* "in any ocean of the globe." Perkins included a color brochure of the boat, just in case O'Brian didn't understand the magnitude of the offer. After making sure that Perkins wasn't a stalker straight out of a Stephen King tale, O'Brian in a handwritten note replied back from his home in the southwest of France, "I accept your kind invitation with perhaps obscene haste."

Perkins agreed to go along on the Mediterranean holiday, for what he assumed would be a sublime time spent with a nautical laureate. After all, O'Brian had written well-paced seafaring sagas that always got the details right about sailing—he knew a scupper from a schooner, a jib from a jibe. Unfortunately, as an incredulous Perkins recounted it, O'Brian was clueless how a sailboat actually worked. When he wasn't steering recklessly or causing more than 10,000 square feet of sail to flail about, he was either drunk or seasick (or both). And he had no idea about time or speed. One day, Perkins asked what port he'd like to have dinner in that night. "Istanbul!" O'Brian answered. Trouble was, the city was 1,000 miles away and *Andromeda* at best could only do fourteen knots. "Clearly," Perkins concluded, "Patrick O'Brian had never been aboard a sailboat in his life." Or he couldn't do the arithmetic.

On *Andromeda*'s first trans-Atlantic voyage, in the fall of 1991, Perkins almost got himself caught in the infamous "perfect storm." Westward bound, from Bergen, Norway, to Newport, Rhode Island, he and his crew knew they were headed into a vicious region of low pressure near the Grand Banks off Newfoundland, with continuous fifty-five-knot winds and waves the size of buildings. Perkins could try to skirt by it, or turn around and avoid it entirely. He chose the former—in part because he had a business meeting to make in New York City—and endured a horrific ride through violent seas and shrieking winds, but lucked out. Had he chosen otherwise, he would not have been far enough in front of the weather system that then merged with another low-pressure area to become a once-in-a-century storm. Perkins still wonders if he and the boat would have survived. "It was the time I learned what 'green water' meant," said Vanni Marchini, a Perini Navi production manager who went on that trip across the North Atlantic and later supervised the *Maltese Falcon* project in Turkey. "For days, the waves were solid walls of green," which submerged the leeward rail and sometimes swirled over the bow itself. In heavy weather and grim skies, green was the color that the seas took on; the inviting blue of the ocean disappeared. Green was bad luck on a boat; at Perini Navi, it's taboo as a boat color. On *Andromeda*'s bridge, Marchini kept crossing himself, as if he were in confession. "Vanni!" Perkins asked him, "please stop doing that."

————

With *Andromeda*, Perkins had sailed far from the madding crowds, seeing icebergs in the Antarctic and making it to the Arctic Ocean above Scandinavia. Gerd was Norwegian and loved being on that latter trip. Yet Perkins was also sailing in 1994—without her—when he was finishing his trip around the world and received an emergency message that Gerd's lymphoma was back. He completed that last bittersweet leg into Viareggio, Italy, and then rushed home to California. When she'd been diagnosed with cancer four years earlier, Perkins largely withdrew from KP and dedicated himself to getting her well, through conventional and experimental therapies he helped find for her through his biotech connections. He talked to physicians in the United States, Europe, and elsewhere. He took her to the Dana-Farber Cancer Institute in Boston and the Fred Hutchinson Cancer Research Center in Seattle, and to the Stanford University Medical Center in Silicon Valley. He visited drug labs inside Genentech, the company he had helped to create. He knew physics, but biology and chemistry were a bit of black magic to him, and it took some time for him to learn what he thought he needed to know. Still, when she was in a long remission, he and she thought they had the disease beaten.

So when the call came aboard *Andromeda,* he was crushed. Not only had his efforts apparently failed, he wasn't at home with her but was in the middle of the Mediterranean. After he landed in San Francisco, he drove straight to the hospital at Stanford to see Gerd. A week or so later, he brought her back to Marin and in seven weeks she was gone, at the age of sixty-one. Tom Perkins—scientist, researcher, entrepreneur, builder, fixer—couldn't master cancer and was powerless to prevent the death of the love of his life. He had lost her. And he had lost his conviction that he could control just about anything. Even years later, he believed that if Gerd had gone through a different treatment regimen—a bone-marrow transplant sooner, different chemotherapy, whatever—she might have been saved. Perkins's self-confidence—what had guided him so well and led to extraordinary professional success—was now just generating rumination and personal recrimination. If only, if only—he still appeared convinced that he had the answers. They simply came too late.

He could have kept *Andromeda* for the rest of his sailing life, but she contained too many memories—250,000 sea miles worth of them, with the specter of Gerd hanging over him like Edgar Allan Poe's "Annabel Lee." Perkins held on to *Andromeda* for seven more years—until the *Maltese Falcon* was well underway. When he married Danielle Steel in 1998 (her fifth husband), they did spend time on *Andromeda.* But without Gerd, it wasn't like the *Andromeda* of the earlier days. And Perkins had another boat to fall in love with.

After Gerd's death, Perkins looked for immediate projects to deflect his grief. He knew of the renown of the naval architect Nat Herreshoff, who had built five victorious America's Cup defenders from the late 1800s to the early 1900s. When Perkins learned that Herreshoff's *Mariette* was for sale, he right away flew to Naples, Italy, to see her. He bought the storied schooner soon thereafter for about $6 million, and in early 1995 threw himself into restoring the eighty-year-old yacht. Working with Herreshoff's original drawings that were kept at MIT, Perkins retrofitted *Mariette* with new spars, sails, engines, hydraulics, and electronics. In his old car shop in Marin County, he even made the gold leaf trim himself. It was in the fall of 1995, after a summer of successful racing, that he had the accident at the Nioulargue regatta, and the subsequent manslaughter trial and conviction.

As a companion to *Mariette*, Perkins bought yet another yacht—for another $10 million, including the retrofit. The 122-foot *Atlantide* wasn't just a vintage motor-sailor—an unhurried powerboat with a few small sails to provide stability and an aesthetic—but a historically notable one. Perkins may have gotten rich in the universe of high tech, but he invested his passion in artifacts of a lost age. (Even today, he owns no Blackberry, no iPod, and thinks Bluetooth is a relative of Blackbeard the pirate.) "The boat was a shell," Perkins said. "I was buying its history."

The two-masted *Atlantide* was among the largest of the heroic "little ships of Dunkirk" that in 1940 crossed the English Channel to help rescue 340,000 Allied soldiers off the beaches of France. The "Miracle of the Little Ships," made up of hundreds of merchant vessels and pleasure craft manned by amateur yachtsmen, became part of British folklore. *Atlantide*—christened *Caleta* and then renamed years after Dunkirk (but before Perkins's purchase)—had been built for Sir William Burton, who skippered *Shamrock IV*, the British J-class challenger for the 1920 America's Cup. Burton used the diesel-powered yacht as a "gentleman's tender," which followed his sailboats and provided comfortable accommodation for himself and his guests at various regattas. After World War II, she passed through various owners, appeared in the film *Tender Is the Night* in 1962, but eventually wound up in disrepair, in Malta, being used as a platform for diving charters. Before Perkins claimed her, the boat had even ceased flying the illustrious Cross of Saint George from her bow. That flag was reserved for Britain's Admiral of the Fleet. It was against the law for any civilian craft to fly the cross—except a ship that had seen service at Dunkirk.

Perkins found *Atlantide* in Malta in 1997 and acquired her to perform for *Mariette* as she'd done for her original owner—as a support ship. *Atlantide* and *Mariette* were close enough in size that they could be docked alongside each other. Though she had plenty of berths for crew, *Mariette* would never be as comfortable as a motor yacht. When Perkins raced *Mariette*, he and any guests slept on *Atlantide*. And *Atlantide* was the center of nighttime feasts and partying. Perkins's longtime personal chef from California and Plumpton Place—Richard Mondzak—would fly over to Europe for the regattas and cook for a group of seventy-five or more, comprised of crew, girlfriends, and guests.

Over a two-year period of restoration and re-creation in both Malta and Britain, Perkins turned *Atlantide* into a work of Art Deco splendor, complete with exquisite inlays, Macassar marquetry, etched Lalique glass (with a fiber-optic light fitting that would've made Narinder Kapany jealous), chrome detailing, white carpeting and a tiny nine-foot Bugatti tender located atop the boathouse deck. Scrollwork in the guest bathroom was a scaled-down copy of the elevator doors from the Chrysler Building, an icon of Art Deco architecture. In the main saloon, which had the air of a railcar, Perkins hung a painting of a French country home by an Englishman with the initials W.S.C.—better known as Winston Churchill. A gifted Britain-based designer, primarily of yacht interiors, Ken Freivokh did the work on *Atlantide*, establishing him as the person Perkins would turn to for the *Maltese Falcon*.

When *Atlantide* was relaunched with much hoopla in the late summer of 1999 off Gosport, England, she was accompanied by *Mariette* and the second *Andromeda la Dea*—and all of Perkins's nautical friends. Here he was with a personal flotilla of dazzling yachts and talented professional crews: one for cruising the oceans, the other two in tandem for classic racing. What more could he possibly have wanted?

––––––––

From deciding to build *Felicità* simply for his own pleasure, Fabio Perini had become one of the major builders of superyachts in the world—and the leading one for sailboats. His shipyard remains the only yard that both designs and builds most of its yachts; others typically handle construction, but customers hire their own naval architects. By some estimates, Perini Navi was worth as much as $100 million. By the year 2000, it had

KNIFING THROUGH THE WAVES: At 289 feet, the *Maltese Falcon* is the largest privately owned sailboat in the world. Here, in the Mediterranean, the modern clipper yacht romps along at eighteen knots. Her rig carries a phenomenal 25,790 square feet of sail area.

LEGENDARY TEA CLIPPER: The 212-foot *Cutty Sark,* under full sail in a painting by Frederick Tudgay (circa 1872), still lives today—in a dry dock in Greenwich, England.

HIS OTHER BOAT: Tom Perkins's 122-foot motor-sailor *Atlantide* is one of the heroic "little ships

HIS PRIOR BOAT: Before creating the *Maltese Falcon*, Perkins circumnavigated the globe in the 154-foot ketch *Andromeda la Dea*. The yacht also survived the "perfect storm" of 1991.

NOT BIGGER NO 1: It drove Jim Clark nuts that his gaff-rigged schooner *Athena* wasn't quite as long as Perkins's *Maltese Falcon*, unless you included the thirty-three-foot bowsprit.

THE MASTER BUILDER: Fabio Perini, owner of Perini Navi, the Italian shipyard

THE RIG'S CREATOR: Gerry Dijkstra, the Dutch naval architect and ocean racer

TUNNEL VISION: Wind-tunnel testing of a nine-foot model of the *Falcon* at the University of Southampton shows that the revolutionary DynaRig sailing system is aerodynamically sound.

MAKING THE RIG: At the Yildiz Gemi shipyard outside Istanbul, the giant, thin-walled carbon-fiber masts of the *Falcon* come to life. The masts are hollow and will house the sails.

LORD OF THE REALM: Ever the hands-on engineer, Perkins works atop the 192-foot-high mainmast early on the morning after it was first stepped into the *Falcon*.

"BURYING THE RAIL IN THE SEA": Like any sailboat, the *Falcon* heels in a breeze—and in a

A SAIL EMERGES . . . : The *Falcon*'s fifteen Dacron sails unfurl from mandrels inside the hollow masts. Each sail comes out in opposite directions from its vertical centerline.

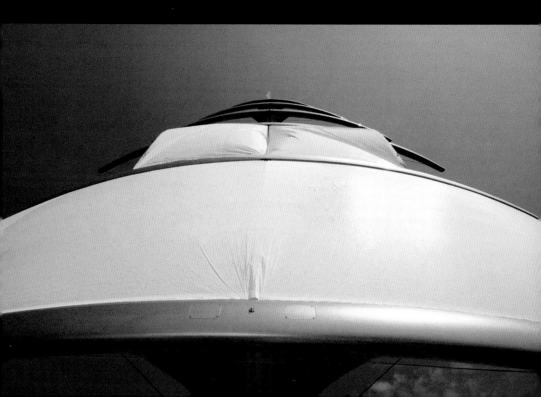

ON THE BRIDGE: From this master station, Perkins can sail the *Falcon*—all by himself, in theory. The center knob is for steering—there is no wheel. The four big, horizontal screens monitor and control the fifteen sails and three rotating masts. In the rear are five screens for navigation, including radar, GPS, and thermal-imaging night-vision.

SEAT OF POWER: This is the sailing-control station, as seen from the opposite view. Behind the high-backed chair is the atrium, where the mainmast enters the boat and descends two levels.

INSIDE THE MACHINE: The mainmast, in the middle of the soaring forty-foot-high atrium, is lit by scores of tiny electro-luminescent lights. From anywhere in the atrium, you can look up two hundred feet or more, through the glass floors and skylights, to see the mainmast and six yards

LUXE ACCOMMODATIONS: The owner's stateroom extends all the way from one side of the *Falcon* to the other. The boat's interior uses a range of exotic materials, including thirty-six varieties of leather. The artwork is large, modern, and weird.

THE WONDERS OF CARBON FIBER: Viewed from the stern, this is the base of the mizzenmast, supported by eight spider-like struts. The couch and tables are for picnics.

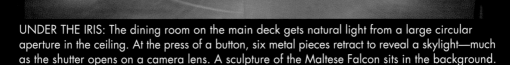

UNDER THE IRIS: The dining room on the main deck gets natural light from a large circular aperture in the ceiling. At the press of a button, six metal pieces retract to reveal a skylight—much as the shutter opens on a camera lens. A sculpture of the Maltese Falcon sits in the background.

DINNER ALFRESCO: In good weather, the aft area of the main deck is the best place to dine. In the background is the indoor/outdoor bar and the saloon, with its large plasma screen TV.

ALL HANDS ON DECK: The *Maltese Falcon* crew perched on the lowest yard of the mainmast, with Perkins in front. One of the sails is furled inside the vertical slot of the mast.

OWNER AND FRIEND: Aboard the *Falcon*, on the lower level of the atrium, Perkins with an Emmanuel Chapalain aluminum sculpture of a menacing shark.

built thirty sailboats. That wasn't a big number, yet it represented the majority of sailing superyachts in existence. Fabio Perini's customers included Rupert Murdoch, the media magnate; Silvio Berlusconi, the on-and-off Italian prime minister and that nation's richest person; the late Prince Rainier of Monaco; and an array of overlords from the worlds of Wall Street, Danish coffee machines, and British kitchen cabinetry. On the wall of his office, Perini kept a huge, beautifully framed sterling-silver map that Perkins made for him, plotting with little ships where in the world all his yard's yachts were.

It was a small fraternity covering the planet's oceans, but the boats tended to congregate in the same nautical neighborhoods of the Mediterranean and Caribbean. The male owners all knew each other and loved to discreetly compare notes on whose vessel was bigger, better, faster, hotter, cooler. Even though the yachts were not designed as racers, they started getting together every two years for a "friendly" weekend regatta off Sardinia, in what became known as the Perini Navi Cup. Because the boats were of different sizes, weights and rigs, they couldn't compete on equal terms, so a panel of neutral racing judges handicapped the fleet in advance: a bigger boat had minutes added to its elapsed time or a smaller boat had minutes deducted. It was supposed to be fair and on that basis it was. But owners, many of whom hadn't raced a day in their lives, went out and hired veterans of America's Cup sailing campaigns—for thousands of dollars. Sure it was "friendly," except when your boat was ahead of mine, not that anybody actually complained too loudly during the cocktail parties and five-course dinners. It was more fun to spend the two years between races quietly carping about the injustices of the handicapping judgments and the outrageous gamesmanship of hiring world-class skippers and tacticians for a weekend of affable competition.

At one point in the 1980s, Perini invited Perkins and a two-time Greek client to each buy 33 percent of Perini Navi. They both declined, as they didn't want to risk their friendship with Perini being spoiled by a business relationship. They also knew that shipyards that built superyachts—notwithstanding the boatloads of money involved—were notorious for going broke. In any event, regardless of where the capital was going to come from, Perini Navi needed to accommodate its growing production requirements. In 1987, it acquired a Turkish shipyard: Yildiz Gemi, in Tuzla, about twenty miles east of the center of Istanbul. Tuzla Bay, connecting to the Sea of Marmara, was a hub of shipbuilding activity for tankers and other large commercial vessels. The area offered abundant land on which to expand. Skilled labor was cheaper than in Italy—perhaps half the hourly wage—and, better yet, Turkish workers put in vast numbers of hours compared to their Italian counterparts, who were used to a workweek of fixed hours and part of the summer off. Perini turned Yildiz Gemi (for "star" and "ship" in Turkish), which had been around for several decades, into one of the largest yards for pleasure boats in the Mediterranean region.

In late 2000, Perkins was in Viareggio for a social visit with Fabio Perini. Meeting in the shipyard's offices, Perkins saw a six-foot-long model of a steel hull that Perini Navi had built on spec and put in an impressive display case—"laying the bait" for some customer at some point, as a Perini executive put it with a smile. Perkins's eyes were immediately drawn to it. There was no rig on the boat, no sails to give it a personality. There wasn't even a way to determine the scale of the hull, yet he knew he had not seen such a Perini before. Its bow was more sharply raked and it appeared to be big. He asked what it was, and he was told it was a 289-foot hull that had been built the prior decade at Yildiz Gemi. It was intended to be for the largest, fastest private sailing yacht in the world.

Why 289 feet? Perini could count, especially when it came to the competition. In 1976, the young French sailor and national hero Alain Colas built a 246-foot, four-masted schooner, *Club Méditerranée*, that he intended to use to break single-handed monohull trans-Atlantic records. Two years later, though, Colas vanished at sea while racing a trimaran. *Club Méditerranée* became *Phocea*, owned by Mouna Ayoub, the ex-wife of a wealthy Saudi businessman. Perini wanted to build a bigger and better yacht than *Phocea*. Since that schooner was seventy-five meters, he arbitrarily added on another ten meters, to which several more meters were added during an initial design study. Not surprisingly, Perini just thought the lines of the boat looked better a bit longer. The total: 88.1 meters, or 289 feet—almost three times the size of a blue whale.

————————

Perkins had previously heard of the hull from acquaintances in the sailing world. He was intrigued enough at the idea of something that big that he contacted Murdoch about it. As head of News Corp., Murdoch had media holdings around the world, yet was not particularly tied into Silicon Valley. He had met Perkins for reasons having nothing to do with high-tech or venture capital. In the mid-1990s, Murdoch—a lifelong sailor—was thinking about buying a sailing superyacht. A mutual friend recommended that he get in touch with Perkins. Their politics were of the same general conservative tilt (Perkins put the framed photo of himself and President Clinton in the guest bathroom of his California home, right beside the toilet-paper dispenser so you couldn't miss it). They both respected the other's success, and they both liked to sail. Murdoch flew to San Francisco, where the second *Andromeda* was at the time. They spent a day on the bay together, and it cemented a personal and business relationship. Perkins told Murdoch of Perini Navi, and Murdoch went on to build *Morning Glory*, a 158-foot sister ship to *Andromeda*. At Murdoch's invitation, Perkins then joined the News Corp. board of directors. In 2006, Murdoch graduated to *Rosehearty*, a 184-foot ketch. It was a deal that Perkins put together impulsively. Murdoch was prowling about for a bigger boat, Perkins called Perini and said to come to New York City with blueprints to see Murdoch that weekend, and that was that. The whole sailing thing worked out very nicely for Perkins and Murdoch.

When Perkins learned of the hull at Yildiz Gemi, he wrote to Murdoch, suggesting he consider buying it. That would give Murdoch the honor of owning the world's largest pleasure sailing craft. Though Murdoch had just lost *Morning Glory* in a divorce, he had no interest. "But you, Tom, you should buy it," he recommended back. Perkins's longtime assistant, Kathy Jewett, had told him the same thing when she typed the letter to Murdoch. Perkins's reaction was, "That's ridiculous. How many boats does one need? I have three!" Little did Murdoch think at the time of his suggestion that he was going to spend the next five years hearing about the boat at board meetings and social get-togethers. "Tom's a fanatic," he said with the admiration and horror of a fellow yacht owner. "Three-hundred feet is fine for a yacht—just not for me."

Jewett, who kept Perkins's books and knew the financial details of his business affairs and hobbies better than he did, knew 289 feet was big for a boat, even for her boss. But it didn't really hit home until a few years later when she was sitting in the Kleiner Perkins box at a San Francisco '49ers preseason game. She gazed down at the field and thought to herself, "eighty-eight meters . . . multiply by about three." Then it dawned on her, "This boat is the length of the football field."

Perini had done the monster-sized hull on spec, during a lull in business. He hoped it would launch a new era in the superyacht industry. But after a prospective French buyer got cold feet, the boat languished for years, at a dock in Tuzla. Perini's instincts were correct as usual—they were just early. Sailboats had gone through the 100-foot barrier, then 150 feet. Powerboats were even larger. So Perini wondered—why not pass right through 200 and 250 feet, and build close to 300 feet? He might have imagined going even larger. But construction sheds were only so big and economics were relevant, even for superrich mariners. In the superyacht industry, the accountants say cost is roughly proportional to the cube of the length, so if your 300-footer was $100 million, then a 350-footer would cost nearly $160 million—a rather large increase*. So Perini learned that the one-upmanship market wasn't quite ready for a sailing superyacht.

Motorboats were another matter. The so-called megayacht and hyperyacht and gigayacht—as size increased, so, too, did the need to dream up fanciful terminology—were especially appealing to motorboat owners, all the more so if they didn't play golf. At the turn of the century,

---

* For those doing the math at home: 300 cubed is 27 million and 350 cubed is 42.9 million. The ratio is 1.59, which, multiplied by $100 million, yields $159 million.

billionaires like Paul Allen (cofounder of Microsoft), Roman Abramovich (the oilman and Russia's richest citizen), and the crown prince of Saudi Arabia had motor yachts surpassing four hundred feet. The trade publications that catered to this rarefied realm loved to egg on the haves and the have-mores. "SIZE MATTERS!" announced a typical yacht advertisement in *Arabian Knight* magazine ("circulated by hand" to the leaders of the Arab world, "individuals who shape the destiny of the Middle East through their enormous power and fabulous wealth," and presumably their really big boats, too). Bernie Ebbers, the disgraced former CEO of WorldCom (now doing twenty-five years in federal prison), glibly anointed his yacht *Aquasition*. "Grow, grow, grow your boat," went the tune sung by shipyards.

One of the ironies of owning a giant yacht, though, was that most of their owners maintained a degree of secrecy about them, as if such a vessel's mystique were enhanced by discretion. Some shipyards—building boats so big that they could be seen on a satellite photo—nonetheless liked to deny they were building anything big, or at least to refuse to acknowledge who commissioned the yacht, even if that fact was well-known in the industry. The yachting trade press, as well as such cable outlets as the Travel Channel, were complicit in the arrangement, honoring owners' requests for anonymity in return for photographs or video tours of a boat's extravagant interiors. The secrecy didn't stop an energetic network of yachting paparazzi from tracking the location of the three thousand superyachts across the globe. At yachtspotter.com, for example, the rabble can see where the boats have been—exterior photographs included—and if there are rumors about where they're heading. Discussion boards at other Web sites pore over sales, celebrity spottings, and everybody's favorite—mishaps like groundings, collisions, and crew misadventures with the local constable at seaside bars.

Megayachts and gigayachts (and, alas, McYachts) came with such equipage as twin helipads (with twin helicopters—one for the owner and the other for everybody else), a glass-bottom indoor pool covering a coral-reef aquarium, a basketball court, a recording studio, and concert space for 260; underwater speakers; multiple grand pianos, a bowling alley, a hair salon, scooters, and transportable Land Rovers (stored in a "garage" that opened from the stern of the boat), in addition to a de rigueur movie theater, spa, Jacuzzis, and an assortment of watercraft toys, ranging from a Sunfish to windsurfers to jet skis to a rocketboat that went 117 knots. With the right naval architecture, buoyancy trumps gravity every time. Yet looking at the things that these boats took aboard, you had to wonder if the laws of nature just might want to play a trick or two on all this excess. Abramovich's *Pelorus* even had a submarine for locating mines, along with a missile-detection system. Another yacht reportedly carried surface-to-air missiles. The 452-foot powerboat of Oracle CEO Larry Ellison had eighty-two rooms. If you're counting how many satellite-connected flat-screen TVs are onboard, you're just gawking.

Yacht-building is surely a growth industry. As of early 2007, according to an annual survey by *ShowBoats International* magazine, there were current orders in for 777 luxury yachts that were eighty feet and longer. Those yachts—most of them motorboats—totaled 94,404 linear feet, a 15 percent increase from 2006. Pity the overburdened custom shipbuilders, one of whom complained that the wealthy population was "increasing faster than the shipyards can build the boats they want." For the billionaires who wanted the boats, price was beside the point. Some were happy to bargain—to them, dealmaking was Pavlovian—but most just wanted the boat. Abramovich supposedly bought *Pelorus* four weeks before it was completed—for someone else; as the story goes, that initial buyer couldn't resist getting double from Abramovich what the buyer had agreed to pay the shipyard.

No sailboat had enough volume or width to carry all those trappings—a sailboat had to be able to *sail*, so it couldn't be a floating mansion—and many a mogul wanted it all. That's why only about 5 percent of the world's superyachts were sailboats. A colossal yacht, Microsoft's Charles Simonyi told a financial publication, was "the closest a commoner can get to sovereignty." It was no surprise, then, that Simonyi's gray

233-foot *Skat* looked rather like a warship (with frequent sightings of that commander of taste, Martha Stewart, aboard, according to yachtspotter. com). But if a sailboat got big enough, it could have plenty of room for some of the accouterments that the stinkpots had. And the longer the hull, all things equal, the faster a sailboat could go.

That was just one of the laws of physics, a subject Perkins understood almost as well as the laws of economics. A nonplaning sailboat simply could not escape the wave it made as it went. As the bow cut through the water, it generated a bow wave (that's also the photogenic V-shaped wake of white, foamy water spreading out and trailing from the bow of any moving vessel). That wave got bigger as the boat went faster, no matter how sharp the bow of the boat (though a sharp, knifelike bow, like that on a clipper ship, reduced the size of the bow wave). To climb up and out of that trough took energy that a sailboat might otherwise be using to go faster, but the boat couldn't go faster while it was impeded by the bow wave. The size of the bow wave was primarily a function of the boat's speed, though sea depth and waves caused by the wind also affected it. But the longer a sailboat was, the faster it could go before the bow wave got in the way of the hull. The main reason that many sailboats had a sloped transom—the surface at the stern of the boat—was to make the boat's waterline longer and therefore better able to handle the bow wave.

An influential nineteenth century British naval architect named William Froude studied water resistance and was the first to devise a formula for determining a vessel's "hull speed." That was roughly the maximum velocity a nonplaning sailboat could attain, unless one applied logarithmically greater amounts of power. A multihulled sailing vessel like a catamaran could go faster because it could skim over the surface and partially climb out of the hill that the wave was creating. A speedboat of course could go virtually as fast as it wanted as it planed over the water, as long as its engine provided enough juice. Even a traditional monohull could exceed its hull speed by sailing down the back of a wave, or by getting extra power in a strong wind and staying upright because of a sufficiently deep and heavy keel. In these conditions, the boat tended to rise above its bow wave and seemingly defy Froude's calculation. But these were exceptions that were temporary.

Froude's Law states that maximum hull speed is the square root of a boat's length at the waterline multiplied by 1.34.* A heavy hull, or one with an inefficient shape, obviously couldn't approach that full hull speed. And the benefits of length were diminishing, since the relationship between speed and length is not linear, but based on a square root. But even so, Froude's Law was clear: the longer a boat, the faster it can go. Thus, a 100-foot sailboat in theory could go no faster than 13.4 knots; a 200-foot sailboat, no faster than 19 knots; and a 300-foot sailboat, 23.2 knots. (The Perini hull at the waterline was 257 feet, so its maximum theoretical speed was 21.5 knots—three knots *faster* than its two diesel engines could make it go.)

For skippers who wanted speed—while also having a hull with creature comforts and that could be used for cruising—going long was the only way. For length also produces stability—a boat's ability to stay upright. In fact, stability goes up exponentially—to the fourth power—as the waterline length increases. Assuming all other measurements equal, if you double the length, then stability goes up sixteen times (2 × 2 × 2 × 2). Stability and length are among the few relationships in physics subject to the fourth power. The tendency to heel increases with length as well, but only by the *third* power: if you double waterline length, the heeling force goes up only eight times (2 × 2 × 2). So the benefits of length still predominate, as the increase in stability is greater than the tendency to heel. God must be a sailor.

For multiple reasons, then, in the world of superyachts, a longer sail-

---

* The 1.34 multiplier isn't absolute. Other factors, like water density and salinity and temperature, as well as air temperature and humidity, might affect it slightly. So, too, do overall sea conditions, the smoothness of a hull, and the weight distribution within a hull. Naval architects after Froude said the multiplier varied between 1.15 and 1.45, though the convention remains 1.34. It is a guideline rather than an absolute. It doesn't work at all when one plugs into the formula a tiny vessel or one of infinite length. Contrary to Froude's Law, a vessel that is a few inches long still moves. And an infinitely long vessel does not go infinitely fast.

boat made particular sense. "Bigger was better," Perkins would say over and over. Ego and utility came together. With *Atlantide*, as with *Mariette*, Perkins was buying not only two yachts, but their heritage. With the *Maltese Falcon*, he aimed to create a new one.

———

Standing in Perini Navi's Italian offices that day in late 2000—six years after Gerd died—as he pawed over the model 289-foot hull, Perkins's mind raced with the possibilities. Such a large yacht was nearly double the length of his second *Andromeda la Dea* ketch and seventy-seven feet longer than the *Cutty Sark*, the great old tea clipper that many said was the most famous ship in the world. The hull was not only long, but relatively narrow at forty-two feet; a motor-sailor of this length could easily have a fifty-foot beam. And the hull sat low in the water, with a relatively modest freeboard (the amount of hull between the waterline and the deck); indeed, its ratio of freeboard to overall length was similar to *Mariette*'s. In a word, the 289-foot boat was sleek. "That hull would look really good with a clipper rig," Perkins remarked to Ragnetti, the Perini Navi CEO.

Perkins's thoughts necessarily turned to Gerd. On expeditions to the Perini shipyard during the 1980s, in pursuit of bigger and better yachts, Gerd was often along. Perkins loved to share his enthusiasm with her, and she obliged him, even if with a bit of teasing. "What would Gerd think of this?" Perkins wondered to himself, as he examined the model. It was a question of curiosity—would he be taking on such a project if she were still alive?—but it was also a lament that she was not there.

After first deciding the name *Atlantic Cloud* evoked a bygone era of tradewind flyers, Perkins came up with the notion of the *Maltese Falcon*. For tax and logistical reasons, he planned to make Malta the eventual home port for his next yacht, having already bought a permanent tie-up in a harbor full of million-dollar berths. (It was another of Perkins's great deals: he beat to the punch the Aga Khan IV, who was buying up virtually all the other berths. So even though the *Falcon* these days continues to fly the flag of the Cayman Islands, the Malta berth earns rental income and appreciates in value.) And since a falcon was the symbol of Malta and would make a nice sail logo, the name worked well. It didn't hurt that Dashiell Hammett's detective novel also gave him distinctive names for his tenders and toys: *Brigid O'Shaughnessy* and *Iva Archer*, the tenders; *Effie Perine*, the man-overboard rescue boat; *Wilmer* and *Joel*, the jet skis; and *Fat Man*, the Hobie Cat catamaran.

The Perini hull was supposed to have had a more traditional rig—a three-masted "sketch," which was half schooner and half ketch. The one prospective buyer, a decade earlier, had no problem with that rig. He had backed out because he thought 289 feet was just too big—a hundred feet larger than any other private sailing yacht at the time. Perini Navi built only fore-and-aft rigged yachts. Square-rigged sails had never occurred to Ragnetti or Perini or anyone else at the yard. But the boat was just sitting around; Perkins and Perini were pals; and Ragnetti agreed to go outside the company and consult a range of naval architects for ideas. "Tom wanted something different for this unbelievably large hull, something that nobody had seen before," Ragnetti recalled. "We knew that clippers were a throwback, but beyond that, we didn't really know what Tom wanted."

Neither did he.

SIX

Ego Trips

Tom Perkins wasn't the only potentate to be thinking big as the millennium arrived. Sailing yachts of 150 feet or so were no longer novel. Perini Navi had built the majority of them—nearly two dozen. Strong, light materials like carbon fiber and other composites were becoming less expensive, and designers were becoming more comfortable working with them in constructing masts and boats. It was logical to assume that other people with overflowing bank accounts and technical backgrounds might be dreaming of the first sailing megayacht. There were two such individuals—like Perkins, both Americans: Jim Clark, the three-time high-tech winner, and Joe Vittoria, who made a fortune more traditionally, as head of Avis Rent a Car when he led a leveraged buyout of the company in the 1980s that produced the largest ESOP (employee stock ownership plan) of the time. Clark had in mind an enormous gaff-rigged schooner, which he'd name *Athena*. Vittoria planned to build the largest sloop in history, *Mirabella V*, with a mast that was higher than the boat was long; to emphasize the point, he specifically wanted a mast that was at least 20 percent taller than the boat's length. Perkins, Vittoria and Clark were the Vanderbilts of the new yachting age—each with a very different boat, each with his own exemplar of conspicuous construction.

A decade younger than Perkins and twice as wealthy, Clark was the enfant terrible hero of Silicon Valley colorfully profiled in Michael Lewis's 1999 bestseller, *The New New Thing.* With degrees in physics and computer science, Clark had launched three billion-dollar start-ups—Netscape, Silicon Graphics and Healtheon (now WebMD)—and made hundreds of millions from them. Netscape was his grand-slam home run. Its creation of an easy-to-use browser to surf the nascent World Wide Web, for better and for worse triggered the dot-com boom, and its initial public offering in 1995 gave rise to an orgy of capitalism. Clark was the No. 1 prospector. He and Perkins knew each other from the tightly knit Valley milieu, especially since Kleiner Perkins had backed Clark's start-ups. They even occasionally wound up on the speaking circuit together. When Perkins gave his speech in New Zealand in 2000 that seemed to presage the collapse of dot-com stocks, he happened to follow Clark to the lectern. Clark had painted an entirely rosy picture of the Internet. Everyone in the audience knew the two of them were jousting. Perkins and Clark were well aware of each other's sailing hobby and proclivity for maritime grandiosity. Each was genuinely curious about what the other was doing: how fast their boats went, what exotic island they were visiting, which heads of state they were entertaining, and most important, what they were planning to do with their next yacht, since there always had to be a new new nautical thing.

The only thing funnier than listening to them trying to find out how big the other guy's next boat was going to be was listening to them explain that, no, they weren't that interested in comparisons. When Clark found out that Perkins was considering a replacement for the second *Andromeda la Dea,* he set out to discover how long it would be, for he, too, was shopping for a new boat. When the two next saw each other, Perkins graciously obliged him and let him know that Perkins's next boat was going to be . . . a few feet shorter than it really was. "That's how I made sure the *Maltese Falcon* was going to be longer," Perkins explained with manifest glee.

Clark eventually learned of Perkins's head-fake and went bananas when he next saw Wolter Huisman, the owner of the Dutch shipyard that was building Clark's new boat. Huisman was revered and his family-owned business that dated to 1884 had a reputation as one of the world's finest yachtmakers. Huisman told Clark that *Athena*'s hull was complete and that there was no way to make it longer, unless he wanted a newer new boat before he even took delivery of the merely new one. Clark could console himself only with the fact that if you included his thirty-three-foot stainless-steel bowsprit as part of the length, then his was bigger than anybody else's. Few people included a bowsprit in a boat's length; typically, you measured along the waterline (where the *Falcon* was longer) or, most likely, along the deck, in which case the *Falcon*'s 289 feet was substantially longer than *Athena*'s 262 feet. But is anyone really keeping score on this kind of thing? Well, yes. That's part of the privilege of having the yacht to begin with.

When Clark held a launching party for *Athena* in the fall of 2004, Perkins attended and Clark gave him a personal tour of the boat. Clark was coy. "Tom," he said, "you're building a boat, aren't you?" Perkins, too, was coy: "Why, yes, Jim, I believe I am!" The next morning, Wolter Huisman—several months away from dying of lung cancer—cordially greeted Perkins and lamented, "Oh, Tom, Tom, I can't tell you how much trouble you've given me. All I hear about from 'Yim' is dis *Maltese Falcon*, 'Is it bigger? Is it faster?' "

In contrast to Perkins's mischief and Clark's occasional reaction to it, Vittoria was exquisitely normal. In the battles of nautical egos, he was a relative bystander. While he actually lived down the block from Clark's mansion in Palm Beach, Florida, the two of them had only met briefly, on Clark's yacht, in the nearby harbor, when Vittoria showed up during a public tour one afternoon. Clark obviously knew of *Mirabella V* and Vittoria introduced himself as its owner. But Clark blew him off. Vittoria was barely miffed. That was just his unassuming way. He was proud of his position in the sailing hierarchy. He wanted his boat to be big and had not the least bit of embarrassment that he had devoted a substantial percentage of his net worth to big boats. "My goal was to do something outrageous," he said. It was simply a question of personal style in conveying his chutzpah. Vittoria's style was understated and could rely on humor. When he went on the lecture circuit to talk about *Mirabella V* at yacht clubs and engineering societies, he gave a video presentation about the boat's construction and characteristics. To dramatize her width, he showed a doctored slide of a red double-decker London bus sitting in the depths of the unfinished *Mirabella V* hull: it looked like a minnow in the belly of a whale. (Perkins and Ken Freivokh came up with a similar gag: a slide of the *Maltese Falcon*, with her soaring masts, crashing into the Tower Bridge in London, with the helmsman asking, "What was that sound?")

Yet Vittoria didn't have his own identity wrapped up in a vessel nearly as much as Perkins and Clark. A few years younger than Perkins, he was still married to his first wife; he spoke of his ten grandchildren with the same reverence he had for carbon fiber; and he could imagine a life that didn't revolve around sailboat projects. The latter acknowledgment, urged upon him by his wife Luciana, was one of the reasons for his forty-four-year-long marriage. Sailboats and building them was just the central hobby of his life. He enjoyed the occasional cruise; he wanted extreme comfort on his yachts; and, ever the Avis master of renting, he figured he could have a big boat pay for its own maintenance by chartering it out ten to fifteen weeks a year. With a civil engineering degree from Yale and an MBA from Columbia, Vittoria had spent his professional life working in the mundane world of rental cars. That industry was hardly as exciting as high-tech. In many ways the affable, elegant Vittoria had the most conventional job path of the three yachtsmen—a career born of serendipity and romance rather than the pursuit of wealth.

Like Perkins, Vittoria had grown up near Long Island Sound and, as a teenager, was even more of a sailor, hanging out at the local yacht club, earning cash by scrubbing decks. He was drafted after graduating from business school, but was declared ineligible for military service because of asthma. By then, he was out of synch with the regular recruiting schedule for MBAs, so his father suggested he "go see where you came from"—to meet his relatives in Italy. Vittoria said his father had pretty much been a well-to-do playboy in Italy and that *his* father had sent him to the United

States to get him away from the casinos. Vittoria's father met his wife and raised six children in the suburbs of New York City. It was thus with a certain symmetry that a twenty-five-year-old Vittoria headed off to Italy, was picked up at the airport by his uncle with servants in tow, and taken to a villa on the outskirts of Rome. Vittoria seemed to be picking up where his father had left off a generation earlier.

But Vittoria did little cavorting. A cousin introduced him to a friend—a fellow sailor—who had a sister celebrating her eighteenth birthday. Vittoria went and was immediately taken with the young Luciana. In the early 1960s, they could spend little time alone, and her family deemed Vittoria little more than an unemployed American. She certainly could not return to the United States with him after the summer. He went home and sought a way to make it back to Italy. Hertz gave him a job first in Manhattan and then within the year in Rome. Vittoria and Luciana had written to each other frequently, but he didn't tell her he was returning. One afternoon, he showed up at her family's house and was greeted by Luciana's sister asking, "What are *you* doing here?" To which he announced, "I'm here to marry Luciana!" And so he did, after her family approved upon learning he was making a handsome $8,000 salary. Vittoria stayed in Italy another nine years, then was in Britain for another eleven, winding up running Avis's global operations outside North and South America. Following a falling out with Avis, he tacked back to Hertz, which made him CEO and finally brought him home to the United States. The prior CEO went on to the parent corporation, RCA, but then returned, which sent Vittoria back to Avis, where he made the financial riches that permitted him to get into superyachts.

In his early thirties, he had started small—with a 43-foot sloop. All his boats were sloops—that's what he had sailed at the Sea Cliff Yacht Club on the north shore of Long Island, and that's what he regarded as the most traditional kind of sailboat. For more than a decade, he bought and sold a boat every year, seeming to like the process more than the acquisition. In this admitted love of "projects," he was just like Perkins: tonight's dance was dandy, but tomorrow's courtship promised the possibility of perfect love. By the late 1980s, with millions of dollars available, Vittoria went big, building the first *Mirabella*—for "beautiful vision"—at 131 feet, and then in 1994, *Mirabella III*, at 137 feet. Both were built at a yard in Thailand that he created. (He sold what would've been No. II before it was finished.) At the time of their construction, *Mirabella* and *Mirabella III* were monster sailboats—the largest sloops ever. Although Vittoria and his family used them several weeks a year—never on long voyages, never to the extent Perkins drove his yachts—the *Mirabellas* essentially were charter vessels (rentals by any other name), costing $50,000 a week or more, not including fuel, food, booze, dockage fees, or crew tips. The money paid for the yachts' upkeep and Vittoria believed he was doing something for the good of the sailing fraternity. "It frustrated me how many people built motorboats," he said.

He liked to tell the story about his friend Roger Penske, the race-car driver, who challenged Vittoria to a race between one of the *Mirabellas* and Penske's 153-foot motorboat, *Detroit Eagle*. While Penske's boat could do thirty knots, Vittoria said that in a big wind a *Mirabella* would still prevail. "You'll start out ahead, but won't be able to maintain any speed for more than twenty minutes," he teased Penske. "Your boat will bounce around in the water and your guests will be hanging on for dear life, while mine are drinking tea out of cups." Vittoria sounded like Fabio Perini ragging on stinkpots. To Vittoria, getting prospective customers to consider a sailboat was as much a matter of leisure principle as charter principal. "I wanted the charter business to be successful, but the idea was to get people thinking about sailing. What do you do on a powerboat, other than bring a book and have a drink? There's nothing to do except get to where you're going. On a sailboat, there's action and participation."

But apparently not enough at 131 or 137 feet. By the late 1990s, Vittoria realized that if you wanted to convince people that a sailboat might be as nice as a powerboat, you'd better build one as big. While a decade earlier Perini was interested in making sailing yachts easier to manage, Vittoria mostly just wanted them bigger. He always remembered that on his old smaller yachts, fifty and sixty feet long, women on board complained about having to climb down a ladder to get into the cabin. "Why is this like a submarine?" one asked Vittoria. Early on, Vittoria thought 190 feet would be long enough to make *Mirabella V* sufficiently comfortable. (He skipped over the name *Mirabella IV* because he thought "IV" looked bad in a logo on polo shirts.) Jim Clark had already launched the 155-foot *Hyperion*, then the biggest sloop ever. So if anybody was measuring, 190 feet would trump that. Unfortunately for Vittoria—or fortunately, depending on when one asked him—that length quickly proved to be too small.

The initial reason wasn't the attributes of the sailboat itself, but because of the tender that Vittoria wanted to service it. He knew that the first vessel that any charter guests saw wasn't the yacht, but the little boat that brought them from the dock out to the yacht at anchor. (Most huge sailing yachts had keels that were too deep to permit tying up to a dock; *Mirabella V*'s 150-ton keel was cleverly retractable, allowing the boat's draft to be reduced from thirty-three feet to thirteen, so the boat could at least make it part of the way into shallower harbors. At its maximum depth of thirty-three feet, the keel gave the boat a lower center of gravity that helped the boat heel less, which in turn allowed it to carry more sail area and drive faster.) The "little boat" that Vittoria had in mind for his superyacht was a twenty-nine-foot jet-propelled Hinckley Picnic Boat, an immaculately finished retro runabout. Costing roughly $300,000, the Hinckley was itself a powerboat that most riffraff would be eager to own. Vittoria intended it as the impressive opening act of his weeklong charters.

The only catch was that in order to store the Hinckley on the new sailboat, there had to be a very large "garage" in the stern of the boat. With its own giant vertical hatch, the garage—or "lazarette" in nautical talk—had a track onto which the Hinckley would be pulled in. The garage would also store other toys, such as four Laser sailing dinghies, canoes, kayaks, windsurfers, two small remote-controlled *Mirabella* models, and scuba gear. Given how big "Joe's Garage" had to be, *Mirabella V* had to grow proportionately larger to accommodate the required staterooms, saloons and outer deck areas. As the yacht got longer, Vittoria decided a sauna and gym would be good, too. All this extra room meant the boat would be well over 190 feet. "My other boats are sailboats," he said. "*Mirabella V* would be a resort." Her guests would include the likes of Michael Douglas and Catherine Zeta-Jones, who of course enjoyed the open-air movie theater on the upper deck (the screen was set in the back of a large settee), just a call away from the six-hundred-bottle wine cellar. The initial charter fee was $300,000 a week in high season—not really that much, according to the owner. "There are five hundred or so billionaires in the world," Vittoria often said, "and I only need to rent to fifteen or twenty of them."

By this time, Vittoria had grown weary of building in Thailand. Labor was cheap, but some practices were antiquated. When the early *Mirabellas* were launched, they first had to be towed across a beach by a team of elephants. Craftsmanship wasn't up to the standards of European shipyards, or to the aesthetic requirements of his wife, who was in charge of the interiors. Vittoria said he had nightmares about the protests of Thai workers, "No see! No see!" in response to his complaints about rough edges or marred surfaces. So he found a large public company in Britain, VT Shipbuilding, that was eager to take on an unusual individual product. The company (which used to be called Vosper Thornycroft) was mainly a manufacturer of warships, making minesweepers for the Royal Navy. That was ideal since the minesweepers were made from composite materials and were about the same length as *Mirabella V.* VT had a gap in its production schedule and thought a super sloop would be great publicity. Vittoria saw the chance to negotiate a good deal and he got it—a fixed-price contract, in which the yard had to eat most cost overruns. (*Mirabella V* inevitably had those overruns, in the millions of dollars, and VT took a hit on them.)

To go along with the European yard, he selected a renowned yacht designer and racer named Ron Holland. A New Zealand national who had resettled in Ireland, Holland had designed *Whirlwind XII,* the first Huisman sailing superyacht, as well as boats for Rupert Murdoch, Prince

Rainier, and Edward Heath, the former British prime minister. Holland and Vittoria came up with a blueprint for *Mirabella V* that made her more than 240 feet long. However, because Vittoria assumed that most charter guests would be European, he rounded up the length to the nearest meter, which produced a seventy-five-meter design; guests apparently would be more impressed with seventy-five meters than seventy-four. But *Phocea*— the four-masted schooner owned by Mouna Ayoub that Fabio Perini had wanted to top when he built the *Falcon* hull on spec—was seventy-five meters, too, so Vittoria asked Holland to lengthen the swimming platform at the stern of the boat. At least until Tom Perkins acquired and finished the *Falcon*, that extra eight inches gave *Mirabella V* the momentary distinction of being the largest privately owned sailboat in the world, and even put the boat on the list of the hundred biggest yachts of any kind.

There's a rough rule in the land of personal finance that even tycoons ought to spend no more than 15 percent of their net worth on toys. Perkins seemed to have followed that rule, often noting that "I've made a lot of money and I've spent a lot of money." Vittoria may have been low key, but if he was going to have about 40 percent of his net worth tied up in a rental fleet of superyachts, and to do so to the consternation of his wife, then the flagship might as well command a spot in the *Guinness Book of Records*. At 247 feet, *Mirabella V* was the largest sloop ever built, with the tallest mast—a mind-boggling 292 feet—to match. The single mast was more than 50 percent taller than *Hyperion's*. From the bottom of the fully extended keel to the top of the mast was 330 feet. Vittoria called himself merely passionate about the enterprise. His wife called him crazy.

———————

Meanwhile, Jim Clark was getting ready to build again. After the Netscape stock bonanza, he had commissioned *Hyperion*, a hyper-computerized aluminum sloop he liked to boast he could sail across the Pacific—by remote-control from his laptop at home. He could just as well have named the yacht *Hyperbole*. For while it had satellite communications, and lots of microprocessors to make the shades go down and the music go up at the touch of a button—as the boat sailed merrily along on autopilot—it wasn't realistic that a joystick three thousand miles away could be in control. (Imagine the fun for a would-be hacker.) *Hyperion* was an eye-catching boat that performed well—and predictably, it wasn't adequate for Clark. Before even taking delivery of *Hyperion* in 1998, he was sketching designs of a grander aluminum yacht he'd call *Athena*, after the Greek goddess of war and wisdom. In Clark's mind, the new boat would better serve as a deep-sea diving headquarters; offer two additional staterooms for his star guests, a dedicated theater space, and a library for those who liked to read; and furnish him with an owner's cabin fit for someone of his standing. How many other bigwigs had enough onboard drawer space to sort socks and shirts by color? As much as he drove the Huisman family nuts, Clark was the kind of client a shipbuilder could learn to love, ordering up another boat as the prior one got ready to leave the yard.

As a sloop, *Hyperion* had but one mast. Once Clark wanted to go substantially bigger, he had to graduate to three masts to have sufficient sail area to move at a decent speed. (More than three masts would be unnecessary and impractical.) Apart from the weight of, and stress on, a single mast that was too tall, the bridges of the world placed a limit on mast height. If you wanted to sail your boat between the Atlantic and the Pacific without having to round Cape Horn, you had to go through the Panama Canal and under the Bridge of the Americas. That bridge, connecting the North and South American land masses, had a clearance of 201 feet at high tide. So your mast had to be able to pass beneath it. Likewise, if you wanted to sail under such bridges as the Golden Gate in San Francisco, the Verrazano-Narrows of New York City, or Sydney Harbour in Australia, you couldn't go much above 200 feet. Since Clark intended *Athena* to be used worldwide for oceanographic exploration, he needed to be able to get through the Panama Canal. Unless Clark had any interest in a square-rigger—which he did not—a three-masted ship meant a schooner. After the modern-looking *Hyperion*, Clark liked the idea of a classic yacht—with a handcrafted figurehead on the bow and a backward-sloping fantail transom. Giving it a gaff rig made it look even more traditional; a gaff rig was a pain to operate and made airflow on the sail more tricky, but it did provide the benefit of additional sail area. *Athena*, though, took no chances. Unlike the unprecedented mast that *Mirabella V* had or the *Maltese Falcon*'s unique square-rig, Clark's boat was just another schooner with 192-foot masts, which attempted nothing new in aerodynamics or naval architecture.

Athena had nine sails. There were three mainsails and three jibs—an outer, an inner and a staysail—as well as three topsails above each of the mains. Though the multiple wedding-cake layers of her superstructure made her appear visually top-heavy—much more so than the *Maltese Falcon* or *Mirabella V*, or *Hyperion* for that matter—*Athena* was still stunning to look at, seeming to combine the best of old-world tradition and high-tech modernity.

Yet her appeal for Clark had little to do with sailing. He didn't share Perkins's or Vittoria's fondness for the sea. Apart from providing transportation to, and a base for, diving expeditions, *Athena* was mostly just an entertainment platform. He liked the parties and the gorgeous young specimens that came to them, whether in South Beach or the south of France. Sometimes he even knew the guests; more often they were friends of friends of crew members, whom he had to ask to repeat their names. A Clark gathering meant young beauties dressed in white dresses and slacks, and sporty sunglasses. At Clark's side on *Athena* would be the statuesque Kristy Hinze, the Queensland model less than half his age—the "Aussie Angel" that the Down Under tabloids speculated was connected to Clark's latest divorce. The men and women of *Athena*'s crew were equally tan and handsome, and at parties the stacks of sushi and platters of hors d'oeuvres created an atmosphere of plenty that made Clark a popular host in fashionable ports everywhere.

Attempting to mingle with the guests during an *Athena* party in South Beach to which Clark invited me in the summer of 2006, I thought I was in a *Saturday Night Live* send-up of *Miami Vice*, except I was the only one who thought it was a spoof. Greeting me in Don Johnson garb and bare feet—no shoes on the teak, please—Clark offered me a tour of the boat, but was primarily interested in taking me on the transportable crow's nest that sped 120 feet up *Athena's* aft mast. The crow's nest had rails at waist level, but provided no other protection from falling out. To reach a crow's nest in the olden days, crew had to climb the rigging. On *Athena*, the trip was hydraulic, with the controls in a box with push buttons inside the crow's nest. "What do you think this button does?" he asked me. Before I could even guess, Clark and I were zooming up and down the mast like it was a ride at Disneyland.

The only problem was that Clark—with a glass of Chardonnay in one hand and the control box in the other—was drunk. It was obvious, but he merrily volunteered the news flash as he provided an impromptu play-by-play: *"Wheee!"* I volunteered to throw up on the teak deck if he didn't let me off after the third up-and-down trip. On the way back to the deck, I asked him if he was looking forward to *Athena's* pending voyage through the Panama Canal and on to the South Pacific. "Are you kidding?" he said. "I'm not spending several weeks on that passage. What would I do the whole time? I'll fly out of here Sunday at two a.m., after the party, and then return to the boat once it gets to the Pacific." If your own boat bored you, of course it raised the question of whether you should own it.

Even better for Clark than the vertical joy rides was *Athena's* gadgetry. Not the giant winches that trimmed the sails, or the furling system for the gaff rig, or even the decompression chamber for divers. Or any symbolism in the female figurehead adorning *Athena's* clipper bow, a nod to the nineteenth century. No, Clark's pride and joy were the handheld five-by-seven-inch HP wireless touch-screens that monitored and controlled the ship's accouterments. With a few taps on an LCD monitor—no mouse, no keyboard—Clark could change the music blaring throughout the boat, or order up Beatles in one stateroom and Bach in another, while playing any of hundreds of DVDs on the flat-panel TVs. It all was stored on four terabytes of hard-drive space. Are the shades down in your cabin? Just touch the screen and up they go—or Clark could do so for you from his master-control screen, no matter where he might be on the boat. "This is the only boat in the world that can do this," he boasted. You want privacy and control of the entertainment in your cabin? You're on the wrong yacht. I asked Clark if video cams were next for the guest cabins. He laughed. "Good thought!" he said.

Clark's own suite had a forty-two-inch plasma screen—hidden behind a slide-away oil painting—that could show movies, satellite TV or a giant version of the master controls, just in case he wanted to mess with the music in your stateroom. Clark could also remotely control lighting, heat and air-conditioning, door locks, as well as plot the ship's location on a navigational chart and pull up a digital weather map of where the ship was heading. When he built *Hyperion*, Clark and a cadre of engineers had developed the proprietary software for the screens; on *Athena*, the technology got better. Clark thought the system was hilarious and, to the extent one could speak of Clark having an attention span at all, the HP touch-screens were the closest anything or anybody came to holding it, Kristy Hinze notwithstanding. It was pretty neat, too, according to Clark, that his boat was so big and his gadgetry so pervasive that a lot of onboard communication between himself and his engineers was done by e-mail rather than two-way radio. *Athena*'s crew knew it was working for a man with a few quirks; those crew that remained with him for any length of service tried to keep a sense of humor about the whole thing. They got to see the world, they were paid well, and Clark wasn't on the boat for more than a total of a few months a year anyway. If they had to endure his volatile eccentricities at times, so be it. The crew called their universe "Jim-World."

More than the crow's nest and the electronic toys, Clark had his boat for the image it conveyed—even as he remained awkward on the stages that such cachet opened up. Perkins was a longtime member of the old-guard New York Yacht Club that was founded in 1844. Its landlocked Manhattan headquarters on Forty-fourth Street between Fifth and Sixth Avenues—built on land donated by Commodore J. P. Morgan—was the very heart of yachting, a clique of gentlemen in which Columbus and Magellan probably would not have made the waiting list if they'd been around. After he built *Hyperion*, Clark decided he wanted to be a member. He deserved it, he was entitled. Vittoria and Perkins were both members—Perkins since 1972. Perkins made the calls, got him the introductions, and told him the right clothes to wear. Clark was on the threshold, when it dawned on him he wouldn't have a clue how to comport himself at its Newport regattas, not that he wanted to go to them anyway. "What am I possibly going to do here?" he asked Perkins. Clark didn't become a member of the club. At the launch party for *Athena*, all the guests arrived in white, as the invitations called for; Clark, though, came in a blue suit.

In his prototypical hardscrabble American childhood, Clark had grown up in the Texas panhandle, resentful of others with more money and obsessed with showing everyone he'd do better. He never tired, for example, of noting that Bill Gates "had his way paid to Harvard" and then dropped out. By contrast, Clark had to claw his way through to a doctorate. You didn't have to be Dr. Joyce Brothers to wonder if Clark's striking it rich was a kind of psychological compensation, too. Along with a private jet and an immense house—he had those as well—a megayacht confirmed that you had arrived big-time. That was why Clark was so infuriated when he learned Perkins's *Maltese Falcon* was going to be bigger than *Athena*. When I met Clark on the boat, the first thing he said to me was, "I'm *not* in competition with Tom Perkins's boat." That was also the fourth, ninth, and fifteenth thing he told me. So it occurred to me that Clark might've felt he was in competition with Perkins's boat—though he was never going to allow an actual race between the two. Any time a yachting magazine or yacht club suggested a little sporting contest between all three super sailboats, Clark assiduously declined.

It was a contest he was unlikely to win, taking into account the conventional metric for predicting speed. The *Maltese Falcon* would be longer at the waterline. Froude's Law said the *Falcon*'s maximum hull speed would be theoretically fastest among the trio of sailing megayachts. And if Perkins came up with a square-rig design that worked, the *Falcon* surely would be fastest in fact. Whereas *Hyperion* was a high-performance sailboat, *Athena* just wasn't going to rip through the water. While she was far faster than a plodding motor-sailor and light enough to move well in light air, she simply wasn't a schooner designed to break records. That was never the boat's mission—comfort and looks were—but it roiled Clark that there were going to be two other superyachts, and each of them could claim a superlative that *Athena* didn't have. It was silly, for *Athena* was classically beautiful, and both Perkins and Vittoria said so many times.

––––––––––

The sail area on Clark's boat was limited by having three masts that could only be so tall. That was the price to be paid if he wanted to be able to make it through the Panama Canal. Vittoria felt no such constraints. Indeed, he believed that limiting himself to a mast of 200 or so feet defeated the point of building a superyacht. "Until your mast goes *above* 206 feet, which is the clearance at low tide at the Panama Canal with just the right conditions of the moon, then you're not doing anything new," he said. So his 247-foot *Mirabella V* would have a single mast of 292 feet, the tallest mast ever built—towering over any tanker, or cruise ship, or even a Nimitz-class nuclear-powered aircraft carrier (by 100 feet). It wouldn't come close to making it through the Panama Canal (or under any major bridge in the world). That was fine with Vittoria, who intended the boat be sailed only in the Mediterranean and Caribbean, which were the most popular locations for charter guests. It wasn't hubris on Vittoria's part to build such a mast (or a boom that was ninety feet long and more than six feet wide at its midpoint)—it just struck him as a waste of his time to do what Clark was doing. Vittoria thought *Athena* was a beautiful yacht—"too narrow for me" (by nine feet) and "more covered space for my taste"—but he went out of his way not to "knock" it. He had none of Perkins's pugnacity. Yet deep down, Vittoria didn't quite get *Athena*. "Jim Clark is part of the trio," Vittoria said. "But Tom Perkins and I are the pioneers. All Jim did is build another three-masted schooner." In his own quiet way, Vittoria wanted a yacht with a scale that took your breath away. That wasn't hard to do when your mast was going to be taller than most buildings in a small American city. From his perspective, Perkins likened Vittoria's mast to a twenty-pound bumblebee.

When the mast for *Mirabella V* was being installed in late 2003 by a giant crane in Britain, the crew got a call from a nearby vessel in the harbor. On the radio was the captain—of the newly completed *Queen Mary 2*, the largest ocean liner in history. "What in God's name is that?" he asked, as he watched the crane take the piece of carbon fiber and step it into the hull of *Mirabella V.* Vittoria relished these tales of incredulity. "If this boat ever has a collision at sea," cracked an Australian columnist, "it will probably be with an aircraft." Several years later, Vittoria was on the island of Capri, waiting to go out to his boat. Another person was on the dock—Lillian Vernon, the catalog-retailer queen, on her way to her 155-foot charter. When an inflatable raft came by, she mentioned to Vittoria that she was going to get drenched. He offered her a ride on his tender, but Vernon wasn't sure which sailboat was hers in the crowded anchorage—she thought they all appeared alike. "No, they don't," Vittoria told her. "Look at that yacht over there—the one with the red beacon 300 feet in the air." Vernon was amazed. Vittoria was pleased. The light at the top of the mast wasn't there because of commercial aircraft—as some wags suggested—but as a warning to helicopter pilots landing on motorboats like Paul Allen's. *Mirabella V* clearly stood out. Sailing such a boat, however, was going to be a different matter. With giant scale came giant problems.

Vittoria's sails weren't just large—they were staggering. The largest of the three jibs—the UPS, for Utility Power Sail—cost $250,000 and was the largest sail ever made. With 19,730 square feet of synthetic Vectran fiber—a space-age liquid crystal polymer that had been used for the Mars Rover airbags—the UPS could drape a small suburban house. The load on its clew—the aft corner where the UPS attached to the thick sheets controlling its trim—was equivalent to hanging twenty SUVs from that location. The mainsail, made of heavier cloth, was only about 15 percent smaller and weighed more than a ton, so heavy that it couldn't be lifted onto the boat in one piece; final assembly of the mainsail's seven sections was done onboard. The mainsail was also cut in a slightly different way than a true triangle. This "roach cut" near the head gave the sail a rounded outer edge that generated a better flow of air along the sail. The overall effect meant the sloop could roar upwind—with just a few qualifiers. For starters, *Mirabella V* couldn't tack without partially lowering her mainsail—a fairly preposterous process, akin to switching your car engine on and off every time you turned onto a different street. Vittoria said that kind of analogy wasn't fair since his boat was specifically designed for steady, long passages rather than tacking duels during a race. It wasn't fair, he said, to criticize his boat for a use it was never intended for.

Fair enough, but like multimasted vessels, such as schooners and square-riggers, sloops were supposed to be relatively easy vessels to maneuver. Gargantuan sails, no matter how high-tech the materials supporting them, still represented weapons when they came whipping across the deck. The sails were under great stress in a breeze and could tear at any moment—an expensive mishap in the case of the UPS. Vittoria often said that on trans-Atlantic trips—relocating between prime charter territories, based on the season—*Mirabella V* generally used her engines rather than sails, just to avoid possible damage to the UPS.

The mainsail had its own issues. It was sufficiently big that, when fully raised, it couldn't make it around the backstay. So, even in light air, the main had to be dropped twelve feet to get around the stay. The sail was also so tall that it might be subject to contradictory loads—a kind of wind-shear effect. The wind might be moving in one direction along the bottom, wider part of the sail, and moving in another direction near the top. Were the boat to tack under those conditions, the flapping about of a fully deployed mainsail would be especially violent. Assuming there were no problems—though there often were, due to tears and the failure of hardware—it could take *Mirabella V* up to fourteen minutes to complete a single tack—even with an engine on for extra maneuverability, just in case. When the crew did it in fewer than eleven minutes, they considered it an accomplishment. Over the course of a day or two of upwind sailing, that was a lot of time parked in neutral. An America's Cup sloop—admittedly a racing machine, but nonetheless a big one—could tack in a few seconds. "*Mirabella V,*" said Vittoria with understatement, "was not a boat in which you went out for an hour or two of sailing some afternoon." He rightly pointed out that his was a cruising boat, not a racer, so that efficiency wasn't that big a deal.

But even hoisting the mainsail was an adventure. It was done with a push of a button, but many people had to participate—and keep their fingers crossed that the halyard doing the hoisting didn't break (which it did on occasion). In addition to the button-pusher was a deckhand sitting in a canvas chair suspended twenty-five feet up. His job was to spray the track in the mast with a water-detergent mix that served as a lubricant. Someone else watched from the winch room to make sure the sail didn't jam, and several crew members on deck and at the helm monitored the entire production for trouble. If a crew member was too close to the sail as it went up, and it had been raining recently, that person got deluged with water.

It was no wonder that everybody was happiest when Vittoria or Holland the designer weren't on board, which limited the number of times the boat was really going to sail at top speed. Surprisingly, it was Holland who remarked rather wistfully on occasion that he thought *Mirabella V* would never be sailed to its potential. Some of the crew thought the boat was just too frightening—the loads put on her sails and equipment too huge to push to their limits. Sailing to windward in a moderate breeze, for example—with the boat heeled over—meant that the great mast was producing compression forces downward of roughly four hundred tons. The hull wasn't going to break apart—the mast wasn't going to drive through the composite material—but it meant that the strain on stays and shrouds and the mast itself was incredible. Things can break under strain. Big things break with bigger consequences.

Jibing on *Mirabella V* was less complicated that tacking: her insurers simply didn't permit her to alter direction that way. A sloop heading downwind had its boom extended all the way out, on either port or starboard tack. But inasmuch as the wind was directly aft of the boat, a slight wind shift would result in an accidental jibe—with the wind catching the mainsail on its leeward side, and causing the boom and sail to hurtle over to the other side of the boat—from a three o'clock position to nine o'clock, or the other way around. If that wind shift was unexpected and the crew didn't control the jibe, or if the hardware restraining the boom failed, the results could be disastrous. Even though the boom was eight feet above the deck and therefore couldn't hit anyone in the head, an inadvertent jibe could still put such strain on the boom, or stays and shrouds, that they could break or bring the mast down.

Mirabella V's nervous insurers regarded jibing a boom her size as just too risky—to the rig itself and to any humans in the vicinity. Instead, if the boat were running downwind with her sails on starboard tack, she'd have to take the long way around to get the sails on port tack. Instead of a slight turn of the wheel away from the wind—say, from heading north-northeast to north-northwest—the boat had to go all the way into the wind and over to the other tack. So, from a north-northeast heading, the boat turned eastward to southward to westward and finally to north-northwest. Given the ban on jibing, that was the only way to do it. In the case of an accidental jibe, well, that would be a matter for Vittoria and the insurers to work out someday, if need be.

Jibes weren't all that the insurance police prohibited. Most high-performance sloops had a spinnaker—a mammoth lightweight balloon-like foresail that was flown free of the forestay, and was used for running downwind and close to downwind. Spinnakers were typically colorful or, on racing boats, carried corporate insignias or other artwork. On *Mirabella V*, a spinnaker would've been 30,000 square feet—or about the size of Delaware—and the insurance carriers said no way, not that Vittoria protested anyway. The insurers also made Vittoria install software that prevented the boat from heeling over too far. At a fifteen-degree angle of heel, an alarm sounded on the bridge; at twenty degrees, the computers automatically released the winches and the sails were let out, thereby slowing the boat down and reducing the heel. The danger wasn't to the boat itself, but to the people on board. Because *Mirabella V* was so big and wide, there weren't many things to hold on to, as there were on a normal-sized boat. On a sailboat of, say, thirty or sixty feet, heeling at even thirty degrees wasn't a particularly big deal. On a Sunfish, you might be at forty-five degrees, but since you were sitting on the high side and since your feet were braced against the cockpit on the low side, it wasn't particularly terrifying. But any significant heel on a superyacht meant the possibility not only of passengers falling down, but barreling into furniture, walls, or each other. Picture standing in your living room with guests and then your house suddenly leans fifteen degrees to either side. *Mirabella V* was forty-nine feet wide at her widest point—from the windward side to the leeward side was a long way to tumble.

One of the liabilities of being trendsetting is that some folks can't wait to see you fail. *Mirabella V*'s schadenfreude moment happened a few months after her 2004 launch, days before the glamorous Monaco Boat Show. And it happened one summer afternoon with thousands of people watching from the beach and TV cameras rolling. On the French Riviera—three hundred feet offshore from Cap Ferrat and around the bend from Monte Carlo—the boat was at anchor, waiting to pick up Vittoria's daughter and her friends at the dock. In a twenty-two-knot shifting breeze and a building sea, the crew knew it was too close to shore, but the daughter insisted on the boat staying put. The professional captain, Johnno Johnston, was caught between the rocks and a hard place. (Joe Vittoria was not on board.) The thirteen-hundred-pound anchor began to drag and, minutes later, before the crew could get the engines started, *Mirabella V* had drifted on to the reef nearby. It stayed there until the next day, as the crowd swelled, as if waiting for a roadside wreck to burst into flames. The Internet provided its own virtual viewing platform. When Perkins learned from a sailing friend that *Mirabella V* was high and dry, he called his captain aboard *Atlantide*, Justin Christou, whom Perkins knew was only five miles away. "Go see it," Perkins told him. A keen participant on the maritime grapevine, Christou was already on his way.

Within the hour, Perkins had received e-mailed jpegs of the circus at Cap Ferrat.

The boat eventually floated itself at high tide; nobody was hurt (though in chauvinistic seafaring tradition, the women crew members were taken off immediately); the captain soon became the ex-captain; the boat's keel damage was repaired in dry dock, at million-dollar expense; and the insurers raised Vittoria's deductible. A government report concluded that the retractable keel, while allowing *Mirabella V* to moor in shallower water, also altered her center of gravity, which gave her a tendency to swing at anchor. *Mirabella V*'s clever design below the waterline had turned out to be an Achilles keel.

Unfortunately, the greater cost of the Mediterranean mishap was a certain degree of ignominy. Running aground was about the worst embarrassment a yacht could face. Doing so in daylight and in good weather—and with an audience—didn't help. Christou was empathetic, but grateful it wasn't his boat. "About the only piece of advice Tom gave me when I was hired was 'Don't run my boat aground,'" he recalled. "There's so much water out there that you should be able to avoid going where you're not supposed to." A captain on *Mariette* had once hit a reef going into Bermuda at four in the morning. Perkins hit the roof, doing a Rumpelstiltskin hat-dance for the ages. Perkins liked Vittoria a lot, but when discussing *Mirabella V* on the yacht-club after-dinner circuit, he'd sometimes referred to her name in French: "*Mirabella cinq*" as in "descend to the bottom of the sea."

The European press was merciless about the *Mirabella V* grounding, having sport with a rich man and his plaything. OOPS! snickered a front-page headline. MIRABELLA ON THE ROCKS! proclaimed *Boats-Yachts-Marinas* magazine, with the requisite puns about a new drink at cocktail hour. One Agence France-Presse report claimed the crew was "dining" in the cabin when the anchor gave way, which led Vittoria to crack that crew members "eat" but never "dine." Some industry journalists speculated if the incident proved Vittoria had gone too far with *Mirabella V. Lloyd's List*, a publication for the insurance trade, wrote that the episode illustrated the "high risks that come with the new breed"—using new technologies and materials "not commonly in use in commercial shipping." That conventional wisdom, though, was incorrect: the grounding happened not due to the yacht's scale or complexity, but because of plain human error. It was no more an object lesson for megayachts than an accident due to under-inflated tires would be for an Indy 500 car. While Vittoria continued to worry that the biography of the boat would forever be linked to the grounding, he kept his sense of humor over it: the screen-saver on his PC is *Mirabella V* on the rocks.

SEVEN

Going Overboard

Tom Perkins couldn't get the big boat off his mind. A few months after he had seen the model of the 289-foot hull in Perini Navi's offices in Italy, Perkins decided he wanted to visit the real thing in Turkey. In the spring of 2001—more than five years before the *Maltese Falcon* actually went in the water—he was in the eastern Mediterranean with his daughter Elizabeth aboard his motor-sailor *Atlantide*. A side trip to Perini's Yildiz Gemi shipyard outside Istanbul was easy. When he first cast eyes on the hull—which was painted in red primer and sat next to a breakwater in Tuzla Bay—his reaction was immediate. "She's not too big at all!" he told his Perini guide. Elizabeth just shook her head; she was not quite so indulgent of her father's yachting ambitions as Gerd had been. Perkins also was struck by how narrow the hull looked, at forty-two feet—only about four feet wider than a boat a hundred feet shorter would've been. By conventional standards, the hull might have been seven feet wider. Long and narrow, Perkins thought to himself, "this boat is designed for speed." Now, he figured, he would proceed for sure.

He remained fixed on the idea of a clipper. The benefit of a square rig was obvious for a boat of this size. Unlike the fore-and-aft rigs of the sloop *Mirabella V* or the schooner *Athena*, a square rig put more sail area lower down for a given mast height—producing what naval architects called a lower "geometric center of effort" for the boat. Particularly upwind, that meant relatively less heeling force and more driving force. In any sailboat with tens of thousands of square feet of sail area, the loads on the sail went up exponentially with sail area, even more so when compared to other, smaller superyachts; a square rig reduced those forces somewhat. And because a square-rigger used many more sails than a sloop or a schooner, it gave more control to a skipper to adjust sail area or position, in order to take into account wind speed or direction. The disadvantage of a square rig, especially compared to a sloop in light air, was that it simply wasn't going to perform as well to windward.

Over a three-month period, Perini Navi mulled over different rig possibilities for what Perkins demanded had to be a fast ocean-going yacht. An in-house plan revisited the idea of a hybrid ketch-schooner, while two respected outsiders sent in their own ideas. Depending on your perspective, it was either a bit incestuous or a reflection of how few really original superyacht designers there were on the planet. Ron Holland, the designer of the rig on *Mirabella V*, weighed in with a classic square-sail concept, modernized with rooms full of winches to make the boat practical. Gerry Dijkstra, a Dutchman who was doing the rig on *Athena*, sent in something more novel: a German contraption called the DynaRig. He did it on an intellectual lark, never expecting it to become the essence of the *Falcon*.

Like Ron Holland, Dijkstra was a leading naval designer, though he had no engineering or architecture degree. Instead, like Perini, he had developed his skills through a love of sailing and by his own wits. Growing up in the heart of Amsterdam, Dijkstra had wanted to learn to sail as a child, but his parents would have none of it: his father worked in men's fashion. That's what Dijkstra's degree was actually in. He started in the naval business in the late 1960s at the age of twenty-five, as a lowly deckhand on English ships. By the time he was thirty, he was skippering charters in the West Indies, building fishing boats in Indonesia for a United Nations development organization, and then becoming a professional offshore racer out of the Netherlands. He was typically participating in single-handed sailing competitions—a chance, as he put it, to join his love of the sea and technology. It was a challenging life, logistically and emotionally, as he had three young children, who always were waving to him from the dock as he ventured off for weeks at a time for what amounted to sixteen trans-Atlantic runs over the course of a decade. Though these were hardly the days of Magellan or Columbus, the technology of the time did seem long ago: navigating by sextant rather than by global-positioning satellites (GPS), and communications limited to erratic ship-to-shore radio on a device the size of a refrigerator. After a near-catastrophe in an eighty-foot trimaran, in which Dijkstra lost the mast and centerboard, he decided to find a more sedate method of earning a living.

In the months between sailing competitions, Dijkstra figured out how to study naval architecture on his own. Since he had far more experience on the water than most students, it was easy to pick up. And he was lucky to befriend teachers at Delft University, the MIT of the Netherlands. The training was unofficial, but it was more than sufficient to get him started as a designer. Dijkstra kept up his racing credentials, even competing on classic yachts against Perkins's *Mariette*. And he still managed to find time with his family to sail the world. Before going over the DynaRig technology with any audience, he was apt to give you a travelogue of his fifty-five-foot "retirement" sloop, *Bestevaer II*, with a slide show of such favorite remote locales as Spitsbergen, a Norwegian archipelago in the Arctic Ocean. Working on the *Maltese Falcon* project didn't chase Dijkstra into retirement with his wife Loon; he said he'd long been promising her to give up work. Others on the project weren't so sure, bemoaning as they sometimes did about "*Falcon* fatigue" over the next five years. But the project did cause enough of Dijkstra's weathered, curly hair to fall out that he appreciated *Bestevaer* all the more so, even if it was a mere dinghy next to the *Falcon* or *Athena*. (Part of Dijkstra's slide show was his own boat meagerly matched up with the two superyachts he helped to create.) Perkins had the advantage of having his undertaking follow Jim Clark's into Dijkstra's nautical studio. Next to Clark, any client seemed genteel and rational, though if you asked Dijkstra to compare the two tech titans, he'd only crack a tight-lipped smile.

The DynaRig proposed by Dijkstra consisted of three freestanding masts, with five squarish sails on each, but it looked nothing like the clippers of yore. Technically the sails were more in the shape of an isosceles trapezoid—especially so for the sails at the top and bottom of each mast. At the top of the mast, the sail was shorter across its top horizontal edge than across its bottom horizontal edge; at the bottom of the mast, the sail was shorter across its bottom horizontal edge. In that way, the tapering of the sails had the overall effect of making the rig look like three giant quadrilaterals. The geometry was easier to understand if you just looked at a photograph of the boat. In comparison to a conventional clipper, the square rig was there, but in a brand new package—a merger of tradition and technology. This was the sail configuration:

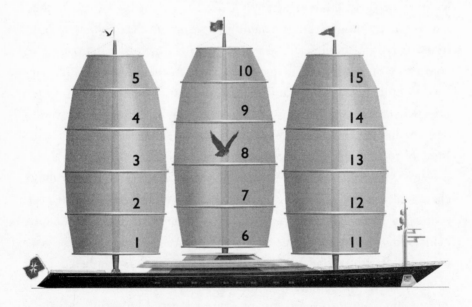

The three sails furthest aloft were the royals: mizzen, main, and fore (sail numbers 5, 10 and 15). Below them were the topgallants or t'gallants (4, 9 and 14), the upper topsails (3, 8 and 13), the lower topsails (2, 7 and

12), and the courses (1, 6 and 11)—except the mizzen course wasn't called that. Instead, sail No. 1 was the crossjack or cro'jack, in part because all terminology is supposed to be sufficiently dense to confuse non-aficionados, but also because historically that sail was sometimes set below the level of the lowest mizzenmast yard, at the level of the flag (or jack).

Had the nineteenth century naval architects Donald McKay or John W. Griffiths seen the *Falcon*, they would've strung up Dijkstra or Perkins from one of the yards. The DynaRig had no rigging, no flailing sheets on deck, and involved no hazardous aerial work by a ship's crew, except to do repairs. Accounts of gallant shipmates—"All lay hands to tackle!" as in Homer's *Odyssey*—would be a thing of the distant past. Instead, the ship would have curved horizontal yards to support and give firm shape to each sail: unlike the old clipper ships, there would be a yard not only on the top edge of the sail, but on the bottom as well. That meant the sails no longer billowed, but were almost rigid and sealed to the aft-sweeping yards. (Some called them yardarms, but those technically meant either end of the yard, which on a DynaRig was a single piece.) Securing the square sails on top and bottom that way also meant there would no longer be any manpower required to keep adjusting the lower edge. The yards—each of which weighed nearly a ton—were not mounted directly onto the mast, but onto large trusses that jutted out six feet forward of the mast.

So unlike on, say, the *Cutty Sark* or *Thermopylae*, there was no longer a series of ropes hanging down from the sails. Nor were there any visible halyards to deploy the sails, or stays or shrouds to brace the giant masts, or the intricate pattern of other lines that rose up from the deck to the yards that distinguished square-riggers. Separated only by the yards supporting them, the five tightly stretched sails on each mast effectively formed a single vertical wing of 8,600 square feet of sail area. Properly trimmed, each set of sails—with their aerodynamically shaped masts and yards—became a tall airfoil. Even though there were five sails supported on each mast, the wind acted as though it was blowing over only one large single sail. As the wind hit the sails, it transferred energy to them. The sails literally slowed the wind down and then—on any point of sail other

than dead downwind—bent the wind along the convex and concave surfaces of the sail, and sent it on to the next set of sails behind. Reducing
the turbulence, or vortices, at a sail's *edge*—where as much as a third of its
drive was sapped—was one of the chief appeals of the new rig. Instead of
twenty sail edges per mast, there were effectively four. When the sails were
put away, they were not bundled up and secured to the yards as was done
on the old clippers, but rolled by machine and stored inside the masts
themselves. In short, there was nothing about the DynaRig that looked
remotely familiar.

When Dijkstra's drawing came into Perini Navi, Giancarlo Ragnetti,
the CEO, took a look and muttered to a colleague, "Whatever that is, it's
not going to sail." Perini Navi knew of Dijkstra's reputation, but had
never worked with him on a project. When Rupert Murdoch first saw the
design, he asked, "Is it going to look so frightening that people won't go
on the boat?"

The DynaRig concept originated with a West German civil engineer
named Wilhelm Prolss and briefly achieved attention during the world
oil crisis in the early 1970s. Prolss was a patent-happy innovator who
made a living designing facilities for German airports. But on the side, he
had long advocated an automated sailing system to supplement engine
power on tankers, freighters, and perhaps cruise ships. He was less an
environmentalist and more an admirer of old sailing ships—he found
them handsome, but an engineering abomination nonetheless. As far
back as the late 1950s, he was proposing his "DynaShip" system as a way
to reduce the cost of seaborne commerce. Prolss's futuristic concept was
to combine the free wind power of old-fashioned square-riggers with the
new technology of microchips emerging in the 1960s. Fuel would be
conserved, without any need to increase the size of crews, for the sails
would be easy to handle. If commercial ships took advantage of trade
winds on the oceans, they would need engines only to maneuver in harbors and to escape the doldrums.

Since no materials existed at the time that could survive the flexing of
a single mast, Prolss's theoretical design, as it developed into the 1960s,

consisted of a large, unwieldy steel tripod that rotated—a tower of three vertical spars onto which the yards were mounted in a fixed position. The sails then were furled and unfurled between the yards. The tripod rotated on a large base, but because it was a tripod, the spars didn't bend. The whole thing, with its different shapes—a three-legged tripod, a circular base, and square sails with rounded edges—looked bizarre, more Rube Goldberg than Germanic utilitarian.

Still, Prolss was a serious scientist and he was convinced his rig was viable. Working with government funds at the Institute of Shipbuilding at the University of Hamburg, Prolss did wind-tunnel tests of a six-masted vessel and maintained that in a decent breeze commercial ships could sail at more than ten knots. But the fuel shortages that came with the 1973 Arab-Israeli conflict were short-lived, so his idea never got beyond the experimental stage and the occasional burst of interest from agencies like the Federal Maritime Administration in the United States. It was just too weird, too unfeasible, and too expensive. It didn't help that other, even stranger ideas were being proposed, including commercial ships powered by windmills or huge kites. And even the notion of reviving cargo sailing ships with traditional designs got pooh-poohed in 1979 after a modest ninety-seven-foot schooner, the *John F. Leavitt*, sank in a Christmas storm on her maiden voyage from Massachusetts to the Caribbean. The *Leavitt*, full of lumber bound for Haiti and the project of a former English professor, was the first commercial schooner built in a half-century and its inglorious demise helped to ensure it was the last.

Various articles in the popular press and technical journals over the next two decades kept the concept of freestanding, rotating masts alive, even as Prolss gave up his vision and faded from public view. The tripod concept evolved into a single pole, based on the emergence of stronger materials. And a U.S. company—based in Silicon Valley of all places—acquired the rights to market the Prolss technology in North and South America. At the end of the day, though, the patents expired and the squadron of DynaShips that Prolss imagined never materialized.

Dijkstra followed the sailing journals, though. When he learned that

Perini Navi and Perkins were considering a square rig for the 289-foot dormant hull, he remembered Prolss and wondered if his idea might be the answer. And he got more excited the more he thought about the DynaRig. In particular, the availability of carbon fiber as a structural material made it realistic to think about a freestanding mast.

Carbon fiber was a miraculous high-tech fabric derived from the most mundane of elements. To make a mast, carbon fiber was first woven into cloth. It was next sandwiched together in a mold—thin black strip upon strip upon strip, with plastic epoxy resin impregnated within layers that ranged from a few dozen to more than a hundred, depending on the required mast thickness. Then the carbon fiber was baked for twenty-four hours at high temperature under pressure to form a composite; because the mast was one piece, the building housing the assembly became the oven. (Silver paint came at the end.) The resulting material was hard, exceptionally strong, incredibly light, and it didn't fatigue like metal because it didn't have the same internal crystalline architecture. Carbon fiber was like a piece of glass that you could ping indefinitely without breaking (unless you did it too hard, but that wasn't a question of fatigue). When you bent and twisted metal, it generated heat—that's why it fatigued over time. By contrast, carbon fiber transferred the forces applied to it almost perfectly—it produced no heat when it was manipulated. It wasn't even noisy when it was bending, twisting, or vibrating. About the only shortcoming to carbon fiber was its frightful expense—orders of magnitude more costly than the aluminum it typically replaced on big yachts. Both *Andromeda la Deas* and *Athena* used aluminum masts.

A metal mast on a DynaRig wouldn't work: it would be too heavy for a yacht to remain stable, and it couldn't withstand repeated bending and twisting. A wooden mast could actually endure more stress, but it also would be too heavy and it wasn't hollow, which would be necessary on an automated rig that had to house cables and electrical wiring, as well as the sails themselves. Carbon laminates, used on military aircraft and sports cars, as well as for fishing rods and golf club shafts, revolutionized how certain high-performance things got made—"the real-world equivalent of

Superman squeezing a lump of coal in his fist to make a diamond," as a *New York Times* auto columnist nicely put it. Perkins wound up buying scores of bobbins of military-grade carbon fiber from Japan—placing the largest single order ever for that material outside the Pentagon. The order was so big that before he could import the carbon fiber to Turkey, he had to convince customs officials he wasn't constructing weapons-grade aircraft. Perkins's three masts contained more carbon fiber than a Stealth bomber or the entire hull of an America's Cup yacht. And he even had enough carbon left over to make the moldings for toilet-paper holders.

At the time Prolss conceived the DynaRig idea, there weren't any superyachts. But Dijkstra recognized that the rig, despite its complexity, seemed to demand a big boat. "In something small," Dijkstra explained, "there were too many moving parts and too many moving components in too small a space." In his view, the DynaRig benefited from being scaled up. Moreover, the bigger the rig, the more sails it could have and the smaller each sail could be. That would reduce the load on any individual sail and it would introduce welcome redundancy. Even if a few sails jammed or otherwise failed, there would be plenty of others to keep propelling the boat. A schooner rig, by contrast, would not only be less manageable, but more dangerous, according to Dijkstra. He thought this, even though he was the designer who built *Athena*, Clark's schooner.

Why would a schooner (or a sloop) be less safe? Any fore-and-aft-rigged yacht on a dead downwind course risked an uncontrolled jibe. That's why *Mirabella V*'s cautious insurers prohibited her from doing even a planned jibe, since the sloop's mast and boom were so gigantic. On a square-rigger, because each side of the mast had an equally sized section of the yard on it, the rig was balanced and jibing wasn't a problem. On the *Maltese Falcon*, it thus wouldn't be an issue, which was a good thing given the importance of jibing on a square-rigger. Because a square-rigger had an identical sail plan on each mast—unlike, say, on a sloop or a ketch—and since the sails were always symmetrical on the mast—as opposed to being all on the port or starboard side of the boat—you never wanted to sail directly downwind in the first place. If you did, the mast and sails furthest aft blocked the wind from reaching the sails on the main- and foremasts.

Drawing up sketches of the rig on the Perini hull, Dijkstra stuck with Prolss's concept of five sails per mast. And given that the height of the masts could be no higher than roughly two hundred feet, the choice of three masts was reached quickly. Fewer than three masts wasn't a choice, because there would be insufficient sail area to reach the boat's speed potential. More than three would aesthetically look too crowded on this hull, and additional masts—all things equal—were never preferable, as they distorted airflow. (That's why a sloop design was ideal, until the boat became too big.) Dijkstra had thought about a fourth mast because he was worried that the loads on three would be too great, but wind-tunnel tests allayed those concerns. He also considered a bowsprit and a few fore-and-aft-rigged jibs. The purpose would have been mostly decorative, as the absence of any sail off the bow of the boat looked odd to some, even if these skeptics couldn't articulate their unease. "Something seems to be missing," they'd say. And many of the old square-riggers did in fact have jibs. In the end, though, Dijkstra and Perkins concluded that those old jibs were largely vestigial and that function needed to prevail over form.

The 192-foot-high carbon-fiber masts were entirely freestanding, with only thin Kevlar wires forming an "X" pattern between each pair of yards—to minimize vibration in the yards. Because they were freestanding, the masts had to be massive and unusually shaped. At the base, they were circular—to allow a connection into the mechanisms that turned them. As the masts rose, they then became wider and elliptical, which was more aerodynamic. And near the top, the masts became tapered for better airflow. At their widest points, the masts were just under six feet in diameter, which made the masts look like silver sequoia trees. A shorter aluminum mast in the bow carried not sails but the boat's navigation and communications equipment, including cell-phone receivers, radar antennae, and satellite domes for television and Internet reception. Because it had no sails, that sculpted "signal" mast didn't rotate, which worked out well since it could then carry forward-directed navigational lights for nighttime—such lights on a rotating mast would've been useless. The signal mast also nicely balanced out the aesthetics of the hull, fulfilling part of the purpose of a jib.

The masts were so massive that even without any sails deployed, it was conceivable that, along with the bare yards, they might by themselves drive the *Falcon* in a strong gale or hurricane. In that kind of storm, a boat needed maneuverability and stability, and the masts and yards alone might substitute as the ultimate stormsail. Yet despite their size, the hollow masts were remarkably thin. Near their bases, where they were subject to the greatest loads, the walls of the masts were five inches thick all the way around. As the masts rose, the thickness of their walls decreased to half an inch—no more than the thickness of sturdy cardboard. That was the marvel of carbon fiber. Dijkstra intended a phenomenal amount of sail area for the rig: 25,790 square feet—enough to cover more than three baseball infields, more sail area than that on three America's Cup racers combined. Froude's Law was Froude's Law, but with that kind of power pack, Dijkstra and Perkins envisioned a boat that could surf downwind or on a reach (where the wind was blowing at a right angle to the boat, directly across her beam) at thirty knots. That was the equivalent of nearly thirty-five miles an hour. If Ferdinand and Isabella had been able to fund such a clipper, we'd be celebrating Columbus Day in September.

———

Fabio Perini—the entrepreneurial wizard who had transformed the paper-making machine industry and then defied the yachting establishment by introducing super sailboats to the world—initially thought the DynaRig was ludicrous. Ragnetti, who already had his reservations about the theoretical rig, showed Perini the various design studies. Perini thought all were plausible, except the DynaRig. "What *is* that?" he asked. Later, he recalled, "I believe in Tom, I always believed in Tom, but I did not believe in this project." In his view, the concept was just "too new"—from the perspectives of both design and execution. "To imagine these giant masts without any supporting stays, and have them be able to rotate, was certainly . . . novel," he said, pausing before he chose his adjective. Though an innovator, Perini prized a certain degree of conformity in the yachts he built. They pretty much looked the same, right down to their dark-blue hulls. Partly because of marketing and partly because of economic efficiencies, Perini liked doing things his way. The DynaRig wasn't the Perini way. Neither was the austere jet-black color Perkins had in mind for the boat.

In his yacht-building business, Perini had met hundreds of prospective clients, all of whom had been sufficiently successful to amass personal fortunes. Yet as much as they understood the laws of economics, they didn't always appreciate the laws of physics, aerodynamics, and hydrodynamics. Perini had heard his share of bizarre ideas. He didn't like to make fun of others' cockamamie schemes, but he also didn't like having his time wasted. Once you got him going on *Spruce Goose* ideas, he reveled in them. There was the guy who wanted to build a 230-foot catamaran (then more than double the length of any catamaran ever). There was someone else whose paper drawings including so many decks and wedding-cake layers that the sailboat would either sink or list over in the slightest breeze. And another person who wanted a colossal mast that would be retractable so it could fit under the Panama Canal bridge. Perini's favorite? The man who insisted on a helipad on his sailboat, notwithstanding the spars and rigging in the way. "We would have meetings with these people," Perini said, "but it's hard to keep a straight face during them. Everyone's *design* project is always perfect. What they don't understand is that making their design come *true* is a different matter."

Perkins, too, had little patience for ideas he regarded as silly—like the eco-friendly, $50-million 190-foot sailboat *Ethereal* that Bill Joy was building. Joy was the software mastermind who cofounded Sun Microsystems and was often referred to as "the Edison of the Internet"; in 2005, he became a partner in Kleiner Perkins Caufield & Byers. By all accounts, *Ethereal* would be a fine boat and Perkins thought the world of Joy's programming expertise. But Joy was a sailing neophyte. And Perkins just loved to talk about Joy's off-the-charts notions. "He sent me a book on

Eskimo kayaks, explaining that the reason they go so fast is that they don't go through the waves, they go with the waves—they are one *with* the waves," Perkins said. Then there was Joy's research into the bumps on the flippers of a humpback whale. Maybe those bumps could be incorporated onto the keel of *Ethereal*? Perkins just laughed, which made one wonder if he realized what folks thought of his own plans. Self-awareness was never his strength. When he periodically likened himself taking on the *Falcon* project to asking the Wright brothers to design a 747 as their first plane, he intended it as a compliment—to himself.

If it weren't Perkins and Dijkstra at the drawing board, Perini might have consigned the DynaRig to his Hall of Fame of foolish ideas. Perini agreed that the new clipper rig could be aerodynamically efficient and incredibly easy to control—if everything worked right and if the vast physical forces could be managed. But normal sailboats had heavily supported masts—by stays fore and aft, and shrouds on the sides. The three fully equipped *Maltese Falcon* masts, weighing twenty-five tons apiece, had to be freestanding for the DynaRig to function. Expecting three columns of carbon fiber, each nearly twenty stories tall, to stand on their own was challenge enough, but adding in multiple moving parts made it even more daunting. "You had to doubt if you could bring everything together to work," Perini said.

Despite his skepticism, Perini was willing to consider the DynaRig concept, along with the more conventional alternatives. The customer had no hesitancy. Perkins thought the DynaRig made complete sense—on paper. So had Bob Swanson's idea for genetic engineering or Jimmy Treybig's plans for a failsafe computer—or, for that matter, Perkins's own foolproof laser, once upon a time. Within hours of seeing Dijkstra's proposal, he wanted to go with it. Perkins was nothing if not obsessive about new ideas. Danielle Steel recalled how he routinely got wrapped up in some science journal or novel. One time, while they were married, he woke up in the middle of the night and couldn't stop talking about photons. "I thought that's what I'd slept on in college," Steel said.

But before Perkins committed to finishing the yacht and building what he intended to be the greatest sailboat ever, he wanted to contain his risk up front. That was the way of the Silicon Valley entrepreneur. He'd spend a chunk at the outset to get the risk out and, as the risk declined, he'd pour in the money, which in this case was going to be roughly 20 percent of his net worth. Even though the *Maltese Falcon* wasn't a business investment and he sought to conquer no market with it—unless you counted trouncing *Mirabella V* or *Athena* in a possible race someday— Perkins concluded that his venture-capital approach still applied.

The irony about Perkins's caution was that Perkins took all kinds of risks on a sailboat. Leaving aside the matter of the 1995 accident at Saint-Tropez, Perkins sometimes had a daredevil approach to the sea—a head-first attitude when tiptoeing might be the better course. On one trip through the Panama Canal, on the second *Andromeda*, Perkins convinced authorities to let him anchor overnight in the lake halfway through. He wanted to go swimming, in and among the trees that had been submerged long ago when the area was flooded to become the canal. When the Panamanian pilot who was guiding *Andromeda* through the canal found out the next day that Perkins and the crew had gone in the water, the pilot told them they were out of their minds because the area was infested with snakes and alligators. The pilot said that never before in his job had he worried that a canal traveler was going to be eaten.

On *Andromeda*'s passage to Antarctica, during her circumnavigation and on a side trip from Cape Horn, Perkins twice asked for trouble. Anchored in a small bay, he and his crew were admiring a flock of penguins frolicking on an iceberg. Several crew members took a tender over to play with them and to see what it was like to stand on a mountain of ice. No sooner did the crew arrive back when the iceberg capsized. If that had happened minutes earlier, the crew would have been doomed—by the time the other *Andromeda* tender was lowered into the water, they would've died of hypothermia. Illustrating the problem of incomplete data, Perkins explained what had happened: "How were we supposed to know that icebergs could capsize?"

Later in the same trip, the boat was on her way back to the Horn, sailing one night through a minefield of icebergs. That wasn't difficult, since *Andromeda* has sophisticated radar: the electronic screen on the bridge nicely showed a measles-like array of blips on the course ahead. Except that all of a sudden Perkins saw a wall of ice off the port side. It was so tall, he said, that he couldn't see its top. Perkins was so startled he didn't even have time to stomp on his hat. He spun the wheel as hard as he could, and the boat missed the iceberg by a matter of feet. The sails came down and that was the end of nighttime iceberg-dodging. Perkins recalled that as the scariest instant of his life. "If we hit it, we were all dead. We were two weeks away from any other boat, out of any radio contact." If he could've reached anybody in such a predicament, Perkins said he'd have had a simple message: "Goodbye." What had gone wrong amid the icebergs? Perkins, the engineer extraordinaire, didn't realize that icebergs sometimes tilt away from radar, so that any radio signals directed at them bounce off and head into outer space. Oops.

EIGHT

Testing the Waters

S till, when it came to potentially spending a sizeable chunk of money on building a boat with an experimental rig, Tom Perkins determined to be risk-averse. There were millions of dollars at stake, not just his own hide. In a different time, this was when Gerd would speak up about family finances. After eliminating any possibility of patent litigation by paying ten thousand dollars to the individuals holding dormant rights to Prolss's designs, Perkins came up with a series of tests—first for the hull, then for the rig. If the testing succeeded, he'd have a reasonably good sense that the *Falcon* was doable; if the tests showed that the hull or the DynaRig were flawed, then he'd be out a few million dollars and could write it off as a lab exercise for the betterment of sailing that just didn't work out. The one thing he would not do was revert to a traditional ketch rig. "Been there, done that," he said. "And I didn't think we could get the performance that I wanted from a conventional rig."

Gerry Dijkstra had devoted himself to the DynaRig idea. It wasn't merely an intellectual exercise—Perkins paid him well. But Dijkstra never seriously believed it would go anywhere. When he found out Perkins wanted to proceed, he told one of his partners, "Is the guy out of his mind?" Perkins asked Dijkstra early on why it was that no one else in more than thirty years had attempted to make the DynaRig happen. "Tom," answered Dijkstra, "you're about to find out."

During nearly two painstaking years—from Delft University in the Netherlands, to the University of Southampton in Britain, to Perini Navi's shipyard in Turkey—Perkins served as his own project manager as he moved the *Falcon* from concept to reality. With Dijkstra, he conceived the tests, and then, based on the results, sought out construction consultants to see what modifications to the hull and rig were possible. Before even confronting the DynaRig, Perkins first wanted to know if he had the right hull. Did it have the proper dimensions for speed? Was steel the correct material? Cost wasn't the issue—a hull would be several million dollars, a small percentage of the overall expense—but time was. A new hull could be built, but it would add a year or two to the project, because Perini Navi would have to start from a piece of paper.

Dijkstra built a simple nine-foot model of the 289-foot hull and put it through weeks of test runs in a towing tank at Delft. Towing tanks had been around since the nineteenth century; Froude himself had taken advantage of them. Using tightly regulated conditions—for the speed of the boat and the characteristics of the waves—the tanks could replicate ocean conditions for a real vessel. While modern tanks often relied on microprocessors to control the variables and lasers to measure the data, they remained low-tech—a few hundred feet long, ten or so feet wide, and a few feet deep. In an age of powerful digital simulations, towing tanks were still the best way to test a hull's design. The counterintuitive thing about computers was that as fast as they were, they weren't very good at definitively taking into account an array of variables in a three-dimensional space. Or at least no programmer had yet developed software that could efficiently do so.

Since the hull was relatively traditional, Dijkstra did not expect unusual results and the tank confirmed that. But the tests nevertheless revealed that, for the rig Perkins planned, there needed to be significant changes. First, because the DynaRig would have far more sail area than a normal fore-and-aft rig permitted (which was what Perini had contemplated for the hull), a different bottom was needed. Though more sail power would make the boat go faster, it also meant the boat would heel more. The force of the wind knocking the boat on her side had to be counteracted by a deeper, heavier keel—like a pendulum, the keel always wanted to be at "rest" in its vertical position and fought back when it was tilted upward as the hull heeled over. The more a boat could be kept upright, the more wind it could utilize to make headway. Some heel was good for a boat, as it reduced the surface area of the vessel in the water— the boat was designed to heel on any point of sail other than going downwind. But too much heel not only slowed the boat down because of water resistance, it also served to dump wind out of the sail as the sail's angle to the wind became too steep.

A boat's keel had long been a baffling, wonderful subject for designers. It was by no means obvious what shape a keel should have: whether it was better concentrated in a narrow deeper section, rather than having a larger wetted surface along the run of the hull's underbody; whether a bulb at the end or the bottom improved hydrodynamics, or whether fins or winglets at the keel's bottom improved performance (it was the controversial winglets on the keel of *Australia II* during the 1983 America's Cup that helped Australia defeat the United States, ending its 132-year winning streak). Keels had two functions: to counterbalance the force of the wind, especially when a boat was beating to windward, but also to act as a wing in the water, much like wings did on an airliner. Hydrodynamics and aerodynamics studied much the same scientific properties, in different mediums. The difference in pressure on a wing's respective sides was what gave lift to the craft attached to it. On a plane, properly tapered horizontal wings made it fly. On a boat, the *vertical* wing—the keel— caused it to drive up toward the wind (just as did the pressure differential on the convex and concave sides of a sail). That's why a boat sailing close to the wind could move forward, instead of just being pushed sideways. Without a keel, no matter how much sail area, a boat would make no headway.

If there was a sweet spot for keel design, no naval architect had claimed to find it. Not even a supercomputer could factor in all permutations to formulate the perfect shape. The dual purpose of a keel is what so tormented designers. There were always tradeoffs between surface area and drag, and between hydrodynamic lift and efficient weight distribution. A computer could tell you precisely what happened on a specific tradeoff—for example, the increase in drag if you increased the keel's surface area—but it had no clue how to start from scratch. Computers were like children. Ask any seven-year-old to choose between vanilla ice cream and a chocolate bar, and they'd do fine. But ask them to arrange dinnertime, set the table, and cook the meal—and you'd starve.

How did the keel tradeoffs work? If you reduced the surface area of the keel—by making the keel deep and narrow, and putting tons of lead at its very bottom—then you reduced drag. But at the same, you risked taking away the keel's hydrodynamic lift that allowed the boat to sail closer to the wind. Then again, if you didn't put a disproportionate amount of ballast down low in the keel, you risked giving the boat too high a center of gravity, which made it heel too easily and hindered speed. You could add ballast to a keel—while maintaining sufficient surface area—but that increased weight. All else being equal, weight was the great enemy of speed. It was a design loop, out of which there was no complete escape. "There was no reason to do this boat except for speed," Perkins said. "It's a thrill to go fast on the water, under the power only of the wind. I like to go fast. Fifteen knots is fun. Twenty and twenty-five knots are even better." Perkins could not wait to see if he could attain such speeds.

The overall shape of the keel also affected the water turbulence created at its edges. Comprehending turbulence—in the water or in the air (around the edges of, and between, sails, as well as near the masts)—continues to be a grail for naval architects, as well as astrophysicists, meteorologists, geologists, and research engineers. For no scientist has ever come up with a definitive way to minimize turbulence. If one wanted to understand aerodynamic drag around a plane or car, or how blood moves through the heart, or why water from a tap can move from a steady stream to a chaotic mess—depending on its flow—then one needed to understand turbulence. Richard Feynman, the Nobel laureate, described turbulence as "the last great unsolved problem of classical physics." The story goes that Albert Einstein—the physicist and well-known sailor—said on his deathbed, "I'm going to ask God two questions: 'Why relativity, and why turbulence?' I'm rather optimistic about getting an answer to the first question."

In short, even naval architects beginning at square one had no certainty they'd devise a keel design that was better than the last one. In the case of the *Falcon*, if Dijkstra wasn't going to design from scratch—thereby defeating the point of using an existing hull—it was readily apparent that he needed to add to its weight and depth to handle more sail area. So the keel's weight was doubled—bringing it up to a total of more than two hundred tons, more than a fully occupied Boeing 737 on takeoff. Dijkstra allocated room in the hull for an additional fifty tons of water ballast, which could be moved around depending on which side the boat was heeling. To prevent intruding into usable hull space, and to give the boat a lower center of gravity, the hull was also made six feet deeper to hold the extra lead.

All that additional ballast, of course, had a potential downside, and Dijkstra had particular worries the yacht might become so heavy it wouldn't be seaworthy. You especially couldn't have a boat that was too heavy too high up; if the boat's center of gravity wasn't low enough, the vessel would be unstable. But you also couldn't have a boat that sat too low in the water—that was just as unsafe, because it could take on too much water when heeling. A sailboat that was heeling too far over was designed to right itself—that's what the keel did, in counterbalancing the force of the wind—unless, while the boat was heeling, too much water washed over the rails and swamped it. Naval architecture, like most structural design, consisted of balancing various factors. And because Perkins's primary goal with the *Falcon* was speed, it was Dijkstra's challenge to push constraints of physics as much as possible. One other change was made to the hull. Since the boat would be faster than originally intended and there would be more strain on steering, the eight-ton, twenty-foot-high rudder was moved aft and extended deeper. That served to make it more powerful, with marginally little effect on drag.

Dijkstra and Perkins did mull over whether steel was optimal for the boat they had in mind. Aluminum (like *Athena's* hull) and a composite of carbon fiber, Kevlar, and high-end fiberglass (like *Mirabella V*) had the advantage of being lighter than steel. Yet hulls made from them—especially large multilevel pleasure boats—required extra fireproofing to satisfy insurance carriers and safety organizations that regulated the construction of big boats that made themselves available for charter. The weight of some aluminum and composite hulls might be nearly doubled by the insulation that regulators demanded for safety. The regulators saw megayachts less as outsized pleasure craft and more as small ships, akin to containerships, tankers, and ocean liners.

More than capsizing from a rogue wave, or sinking in a white squall, or colliding with an iceberg—the fears that Hollywood scriptwriters might imagine—a big yacht's worst enemy was fire. Safety organizations like Britain's MCA—the Maritime and Coastguard Agency—insisted that a megayacht, because it weighed more than the threshold of five hundred tons, be treated like a ship. So its engine room, galley, and bridge had to be able to withstand an hour of intense heat. That surely made sense on the *Queen Mary 2*, where it might take that long to evacuate twenty-five hundred passengers. But on a megayacht carrying a total of no more than, say, twenty crew and fifty daytime guests, the requirements seemed onerous, even ridiculous. Yet the MCA could enforce its will because, without the MCA seal of approval, a megayacht wasn't allowed in any waters of the former British Empire. If you owned a megayacht that you intended to charter in the Mediterranean or Caribbean—or even if you didn't but that you might sell to someone who did—you had to play by MCA rules. The politics of dealing with the MCA—it was an ongoing negotiation between their inspectors and the shipyard—drove both sides mad and since no megayachts were identical, the results often seemed inconsistent.

The MCA wasn't unique. The U.S. Coast Guard and "classification societies," like the American Bureau of Shipping, had comparable rules, yet they weren't controversial because few big boats had an American registry, due to unfavorable tax laws. So it was the MCA that wielded the regulatory hammer.

Yachting safety issues had a rich, infamous history. In the search for speed, radical designs as far back as the late nineteenth century yielded extreme and dangerous boats. In the United States, an incident in 1876 became part of the lore of the foolhardy. The fashion of the time was to make wide boats that had enough presumed stability that they could have shallow drafts. They were called "skimming dishes," favoring simple centerboards over deeper keels. In that Centennial summer, the 140-foot schooner *Mohawk*—one of the largest private yachts of the time—was anchored off Staten Island, her sails set as she prepared to go for an afternoon cruise. She was owned by William Garner, vice commodore of the New York Yacht Club. But a sudden squall hit and capsized the yacht, drowning Garner, his wife, a guest, and the fifteen-year-old cabin boy. "DISASTER IN THE BAY," read the front page of the *New York Times*. In the days that followed came "THE MOHAWK CALAMITY," "THE MATE'S STORY," and "WHAT THE YACHTSMEN SAY." Day after day, the press didn't let up. Stories recounted in grisly detail the "bloated bodies" taken from the water, as crowds looked on from shore. The *Mohawk* captain was arrested and, during a hearing, a mob of five hundred gathered inside the judge's chambers, threatening to "wreak summary vengeance on the object of their wrath." That outcry among the public, as well as the yachting community, influenced safety regulation for generations.

For the trio of twenty-first century sailing megayachts, the MCA fireproofing requirements diluted the weight advantages of *Mirabella V* and *Athena*. Aluminum had a relatively low melting point and carbon fiber burned easily. So Perkins felt vindicated in his preference for steel. (The superstructure, above the deck, was aluminum.) It was a safe choice with the least hassle: it was also the least expensive, easy to repair in case of dents or other damage, and less fussy than composite, even if it was also subject to rust. Had Perkins started from scratch, he might have considered composite for weight reduction, but it certainly wasn't worth losing the construction time.

————

The *Maltese Falcon* hull was simple, compared to what came next. Hulls had been done since forever, and there were only so many variables to play with: they had basic shapes, they all needed rudders and keels, and there was only so much you could screw up. The DynaRig, though, had never been attempted. The trick was to determine if it could work—without building it first. As towing tanks had long existed to test hulls, wind tunnels were the way to analyze sails. The notion of blowing air over and around solid objects under controlled conditions was nothing new. English scientists in the 1700s used whirling arms to measure drag and lift, which led to the first successful unpiloted gliders. But measurements were imprecise. In the late nineteenth century, the Aeronautical Society of Great Britain funded experiments that produced the first enclosed wind tunnel. The breakthrough idea was to move air past stationary models to simulate flight; this allowed measurement to take place under far greater air velocities.

For many America's Cup rigs and other complex sailboats (as well as Formula One cars), the wind tunnel at the engineering school of the University of Southampton in England was preeminent. Dijkstra and Perkins tested the DynaRig there in 2001 and 2002 on another nine-foot model of the *Maltese Falcon*, with masts and sails designed by one of Dijkstra's younger colleagues, Jeroen de Vos. For Perkins, the chance to play at the Southampton facility was part of the fun of making the *Falcon*. For most superyacht owners, testing and technology was a means to an end. For Perkins, tinkering was half the point. Just like spending days and nights upstairs in the laser lab at Berkeley, manipulating the controls in the wind tunnel, and making small modifications in the test model, gave Perkins tremendous satisfaction.

The setup in the wind tunnel was clever. The *Falcon* model wasn't suspended entirely in the air, but rather sat in a little pool of water. It wasn't floating—the water was only there to prevent wind from going underneath the boat and interfering with measurements. The water effectively "sealed" the model, so it acted as if it were in the sea. As in the towing tank, computers kept measurements exact, yet the process was as much art as science. Threads were attached to a range of points on the sail and a smoke wand was used to "visualize" turbulence, but the key measurements were taken from sensors on the model. The sensors gauged stress, sending readings through electrical wires running from the model into the control room, like an EKG machine monitoring a patient. Dijkstra was old school: though aware of the promise of virtual wind tunnels and towing tanks, he was more comfortable in this "real" setting than in front of a computer screen. If Dijkstra made slight changes in sail configuration or the position of the yards, or if he tacked the rig, he could accurately see what that did to airflow and sailing efficiency. These variables in the rig were radio-controlled in the wind tunnel, so there was continuous data as the variables were altered.

Dijkstra and Perkins were especially interested in two forces that airflow would exert on the DynaRig: bending and torque. Masts on all sailboats were subject to stress: a gale blowing against giant pieces of fabric supported on tall columns was going to put strain on any rig. Yet the nature of the DynaRig—freestanding masts that rotated—meant the *Falcon*'s masts would be subject to forces that would be expressed differently. The wind tunnel suggested that a properly constructed carbon-fiber rig would well resist bending and torsional forces. The tunnel also indicated that the DynaRig was sound as an aerodynamic matter, though it offered up an unanticipated finding: the yards needed to have different built-in camber, or curvature. From tip to tip, the yards varied between forty feet and seventy-four feet, depending on their position on the mast—protruding most of them well beyond both sides of the *Falcon* hull. The shorter the yard, the more sharply it had to be curved.

The explanation lay in basic geometry and the preference of the wind. For ideal sailing performance, the wind didn't like to hit a flat surface.

Instead, the best sail shape was slightly curved—what naval architects referred to as a 12 percent arc. (That didn't mean a 12 percent *angle*, but only signified the relationship between the length of the curved yard and the straight-line distance between the two ends of the yard.) Because the *Falcon*'s yards could not all be the same length—since the five sails on each mast were not identical—the yards had different curvatures, even though they all produced that modest 12 percent arc in the sail.

From the data he accumulated, Dijkstra calculated two mathematical coefficients—one for the forces that made the boat move forward (the "driving force") and one for the forces that made it move sideways (the "heeling force"). He then combined those with data from the towing tank to come up with a Velocity Prediction Program, or VPP, for the *Falcon*. VPP, which most large sailboat projects utilized at the design stage, was a complicated computational prediction of a vessel's performance under different wind conditions. It was not a perfect forecaster—naval architects consider accuracy within 5 percent, plus or minus, to be acceptable. That's because so many factors went into VPP, including: sail area and rig dimensions; hull characteristics, like waterline length, overall length, width at the widest point, weight (or, more exactly, the weight of the water that a boat "displaces" when floating), and keel and rudder shape; and such variables as heeling angle and true wind speed (as opposed to apparent wind speed, which took into account a vessel's motion and was greater than the true wind speed when sailing toward the wind). VPP could be expressed in a graphical output called a polar diagram. That showed a boat's maximum theoretical speeds at different wind speeds—and more particularly, a boat's various potential VMGs, for "velocity made good."

If your boat was moving through the water at fifteen knots, but was going more sideways than forward, then you weren't really making much progress. VMG, by quantifying a boat's headway toward a specified destination—it could be a rounding buoy on a race course or the island of Martha's Vineyard as you departed on a cruise from Cape Cod—gave designers and skippers a way to measure a boat's efficiency. Large high-performance sailboats kept various polar diagrams on the bridge—on paper or a computer screen—to measure actual sailing results against. So, VPP and its related VMG estimates were critical at both the design stage and when a boat came into existence.

The trouble with all this was that VPP—because of different input criteria and the limitations of how the sails or rig or hull could in fact be modified—was an inexact business. Even the same data inputs could produce different results depending on what proprietary software was being used. There's a reason that many America's Cup racers, despite tens of millions of dollars put into preconstruction testing by the industry's best naval architects and mathematicians, still turn out to be duds. What looked good in the tank and tunnel—what all the number-crunching promised—didn't always work so well in the water. Yes, a good crew was important—a bad skipper or tactician or winch jockey could add precious seconds to the time it took to tack. But at the end of hundreds of practice runs and qualifying races, most sailors agreed that the America's Cup was won or lost based on hull and rig design. It wasn't that luck ruled the races, but that there was only so much exactitude to be extracted in a nautical laboratory.

Dijkstra was pleased with the VPP data that the wind tunnel and towing tank suggested. For Perkins, it was the first of several eurekas, as he sought to reduce risk and conclude that the expensive task of building the boat and rig should proceed in earnest. But Dijkstra had serious concerns that the boat would really be able to tack, especially in light winds. That was because of the peculiar way the DynaRig worked. Tacking on a conventional fore-and-aft rigged boat like a sloop or schooner was intuitive: you turned the wheel (or, on a small enough vessel, you swung the tiller over), and the boat neatly moved from port or starboard tack to the other. As the boat passed directly into the oncoming wind, the sails were depowered. The sails flapped around as they came across the centerline of the hull. Since the boat was moving forward, it had enough momentum to complete the tack, notwithstanding the temporary loss in thrust from the sails. Even on an old clipper ship, the sails were largely depowered when turning into the wind, since the lower parts of the square sails were not supported by a yard and the sails typically billowed in the breeze.

With the DynaRig, however, there would be no sails flapping around as the mast and yards rotated. The yards at the top and bottom of each sail kept the sail stiff and almost flat. Unlike on any other boat, the sails on a DynaRig would never "luff." Indeed, instead of the sails being depowered as the boat pointed directly into the wind, the sails would be *back*-winded—they would always remain fully loaded, but would drive the boat backward until the boat had made its way around to the opposite tack and proceeded to regain lift to propel it forward. In the old days on the clipper ships, being "taken aback" was dangerous, for the deckhands controlling the sheets could be knocked overboard or, if crew members were aloft, the sails could slam into them. On the *Falcon*, it wasn't that big a deal. If the winds were light—if the boat was moving only a few knots— then it was possible the *Falcon* wouldn't make it through the tack. That wasn't a calamity—it just meant the yacht wasn't going to be able to sail in light air and instead would have to rely on its engines in those conditions. Or the boat would have to change direction not by heading into the wind, but by "wearing round"—turning the bow of the boat away from the wind and then having the rig jibe. It was just the slow, long way around: you went from one o'clock to eleven o'clock not by passing through twelve o'clock but by going in the other direction, past four, six, eight, and ten. Since tacking into the wind was far better, Dijkstra wanted to know at what point the boat stalled because of back-winding. The wind tunnel seemed to conclude that it would be in only very light conditions that the *Falcon* wouldn't be able to fly.

But to dispel his concerns about tacking, after returning to Amsterdam from the wind tunnel testing, Dijkstra built a small radio-controlled sailing model of the *Falcon*. Its sails looked just like what the real yacht would have, it was jet-black of course, and because his young partners couldn't contain themselves, they raced it against another radio-controlled model—a sadly outclassed red sloop. Dijkstra tested it in the canal right outside his Amsterdam office. The model tacked just fine, though it was clear that in very light air the boat was likely to have difficulty. Dijkstra was glad to see that a real-life sailboat, six feet though it was, confirmed the data from the wind tunnel. He even made a video of the model boat and sent it to Perkins, in what became another key moment for the project. Now there was information from two testing regimens that Prolss's DynaRig could be a reality.

———————

The final item on Perkins's initial worry list was the biggest risk: could he come up with a way of easily and repeatedly furling and unfurling the sails—in gales, ice, hot weather, constant salt spray, in all kinds of conditions? All the high-tech testing in the world and all the space-age materials would be beside the point if there weren't some automated way to get the fifteen square sails on and off the yards. Even if Perkins had wanted a crew of thirty or forty to man the sails as if the *Maltese Falcon* were an old-fashioned clipper—and he surely did not, as this was to be a luxury yacht for himself and friends, rather than a dormitory for employees—it didn't matter. The engineering *raison d'être* of the DynaRig was to be automated. If the sails didn't reliably furl and unfurl—if the crew had to climb the masts every time—the rig was a lemon and the project was over.

No computer was going to be able to prove that any hypothetical furling and unfurling method would actually work. So now came the first attempt to build something in three-dimensional space, just like in the good old predigital days. Dijkstra's team spent months trying to figure out how a sail-control system might operate. Would the sails deploy electronically or mechanically? Should they come down from the yards themselves or out of the masts? It was hardly obvious and there were no comparable rigs or products to borrow from. Dijkstra came up with the idea of a system of winches and motors and mandrels that was either elegant or complicated, depending on whom you asked. Either way, there would be dozens of independent mechanized actions to integrate—on each of the three masts, each carrying five sails.

Each winch or motor acted at the push of a button, so the mechanisms were automated in that sense. But they had to work in sync. For starters, it took four motorized winches operating together to separately unfurl each of the fifteen sails onto an upper and lower horizontal yard. The sails were stored on carbon-fiber mandrels housed inside each hollow mast. When the sails were away, they were hidden completely—giving the *Falcon* a look very different from the old clippers, whose furled sails still dominated their profile at anchor. The *Falcon* sails would roll out through feeders onto four recessed tracks—top and bottom, port and starboard—through a narrow four-inch vertical slot in the front of the mast. Each of the four electromechanical winches pulled at the end of a "boltrope" sewn into the edges of the square sail. If one winch pulled too fast or too slow, or if the rope snagged, the whole system shut down and awaited a fix. That way, the sail wouldn't tear or otherwise be damaged. There was no proper in-between point for the sails: to work properly, they had to be fully out.

Since a winch could only pull a rope, rather than push it, an additional device was needed to furl each sail back onto the mandrel inside the mast cavity. That device could be one motor because it was more powerful, needing only to spin a single mandrel, instead of pulling four corners of a sail. Each sail furled from its vertical centerline, folded double onto the mandrel, which acted much like an outsized paper-towel dispenser.

Thus, the fifteen sails needed a total of seventy-five sealed motors to operate—sixty for unfurling and fifteen for furling. The winches were housed on the trusses extending in front of the masts. In addition to reducing the amount of wiring that would've been required had the winches been below deck, the trusses created significant space between the masts and the sails, which allowed for cleaner airflow along the sail surface (almost as if the sail were a flat spinnaker). On other boats, the sails were connected directly to the masts, and the masts disrupted airflow. The machinery for each mast operated autonomously, so a skipper could maneuver one sail at a time per mast. With fifteen sails, and many possible sail combinations for different wind and sea conditions, the *Falcon*'s power plant was flexible—the worse the conditions, the fewer sails you used. Unfurling or furling a sail took about a minute, so if each mast operated one sail simultaneously, fully setting or storing the fifteen sails would take only a total of five minutes. On *Athena* or *Mirabella V*—with fewer sails—the process might take several times that. On *Mariette*, it had taken forty-five minutes to put up the sails and twice that much time to put them away. The *Falcon* also had the benefit of not having to turn into the wind before furling or unfurling sails. On a sloop or schooner, you couldn't attempt to put a sail up while it was being filled with wind—there was just too much stress on the sail and the halyard; and for the same reason, you had to take the wind out of a sail before putting it away.

The sail-control motors were synchronized by computers, but still required the skipper to press buttons to implement a decision. Never could electronics govern the whole process, never could a computer sail the vessel. "No way Bill Gates is controlling my boat," Perkins liked to crack, though the Windows operating system clearly had nothing to do with the *Falcon*'s software. "I don't ever want to have to press CONTROL-ALT-DELETE to restart, in order to make my boat go. Unlike Jim Clark on *Athena*, we'd never let a computer 'close the loop.' The only crashing I want to hear is waves against the bow." Perkins subscribed to Perini's mantra that *"stupido è meglio"*: stupid is better. In the worst case, if the sail-control motors failed, a crew member could climb aloft—either in a bosun's chair or using small handrails on the yards—and manually crank a sail open or closed. (The operation was designed to be done with one hand, thereby honoring the ancient adage for crews: "One hand for yourself and one for the ship.")

———

Each of the three masts also needed devices to make them rotate. That would be controlled by four powerful hydraulic torque motors at the respective bases of the masts. (Remember, the yards themselves never moved—they were rigidly fixed to the masts.) Like the process of sail-furling and -unfurling, rotating the masts required a series of human push-button inputs. The masts on any other sailboat—*Athena*, *Mirabella V*, *Mariette*, the *Cutty Sark*, either *Andromeda la Dea*—were fixed to the hull, braced by stays and shrouds, and held down by their own weight in a "mast step" that essentially was a cup or a tongue-and-groove. These masts had to withstand the tremendous bending and torsional forces unleashed on a vertical pole that carried vast areas of canvas blowing in the wind, but the masts remained where they belonged. That was because all the physical forces on the mast were downward, in effect trying to push the mast through the bottom of the boat. You had to provide sufficient support deep in the hull to bear up to those forces. But at least when the boat was under way, the compression from the stays and shrouds naturally kept the mast in place. The mast needed the compression, for if that force were not present—if the mast went "out of column," in the lingo of naval architecture—it could snap.

By contrast, the *Maltese Falcon* had no stays or shrouds. So she reacted to the force of the wind very differently. When the boat was heeling over, the forces on her masts were partially upward, as well as transverse across her rig. The masts wanted not to drive down through the bottom of the boat, but rather only to topple over and take out anything in their paths. Carbon fiber gave the masts a lot of resiliency to avoid that fate—they were designed to stand up to hurricane-force winds of up to 122 knots and likely could tolerate more—but the masts still had to be secured in some way. Yet paradoxically the masts could not be fixed, since they had to rotate in order to trim the sails and adjust to the wind. That was because the six yards on each mast were permanently attached (unlike on the old clipper ships, where the yards could swivel about an iron ring on the fixed mast, with many deckhands handling the maneuver). The whole point of freestanding masts was that, without stays and shrouds to get in the way, the yards could rotate without restraint, which in turn permitted easy trimming of the sails. The absence of stays and shrouds also meant that the yards could be aligned very closely to the centerline of the hull, which allowed the boat to perform better upwind; on the old clippers, the stays and the shrouds partially blocked the path of the yards.

So, when the *Falcon* was sailing upwind, her masts would rotate so that all six yards per mast were swung around close to the centerline—being nearly parallel with the axis of the keel. When the boat tacked, the mast turned almost a full ninety degrees—from port tack to starboard, or vice versa. When the boat was sailing downwind, the yards would be about perpendicular to the keel. And when the boat was at an angle to the wind, the masts would be adjusted so the yards were in between parallel and perpendicular to the keel. None of the angles of those yards was a new concept—that's what they were on old clippers, and that's what a boom did on a sloop, schooner, ketch, or yawl that had triangular or gaff-shaped sails.

What made the DynaRig system unique was that the entire rig—the masts themselves—moved. The rig was *dynamic*. The wonderful advantage was that the DynaRig eliminated all rigging—including the sheets required to bring a conventional sail from one side of a boat to the other. On a superyacht like *Mirabella V*—less so on *Athena*, because the loads were not as great—those fast-moving sheets, on large winches, imperiled anyone who got in their way, though for most of their run they cleverly traveled under the deck. The sheets also generated intense heat as they moved through their guide blocks. Rising to a temperatures that could top 170 degrees Fahrenheit, as they were subject to forty tons of load, *Mirabella V*'s synthetic sheets might need to be sprayed with water by the crew to prevent them from disintegrating. If one of those exposed ropes broke, the recoil could take off somebody's head. It was just a matter of physics—and sailing an enormous giant boat pushed the physical limits. The safety issue was another potential benefit of the DynaRig on a megayacht.

The masts were kept standing not by stays and shrouds, but by a combination of mechanisms near deck level and below that also allowed the mast to rotate. The base of each twenty-five-ton mast was connected to the hull by means of two huge four-and-a-half-ton steel bearings. The bearings for the two masts most forward on the *Falcon* were just below the deck and deep in the hull. The mizzenmast also had a bearing low down, but because the hull was less deep at this location, the upper bearing had to be above the deck. So, right behind the outside eating area of the main saloon and at eye level, the mizzenmast was supported by eight stylish spider-like struts supporting the bearing. Each of the upper bearings contained spherical cylinders that distributed the loads on the mast. Each of the lower bearings fit into the hydraulic torque motors that were used to rotate the mast.

Because of the upper and lower bearings, rather than a stationary attachment, each mast "floated" in its position. It had to do so to be able to rotate without undue stress on the gear teeth in the bearing mechanisms. Inasmuch as the mast and the hull—the former a vertical piece of carbon fiber, the latter a horizontal mass of steel—were not always in the same position relative to each other when the mast was rotating (particularly when the boat was heeling), the point of attachment had to be flexible. The bearings served that role.

The bearings also had to accommodate the significant flex at the top of the mast—in heavy air, up to ten feet in any direction. Since the base of the mast in the bearing obviously couldn't be allowed to flex that much, the bending and torsional forces had to be transmitted elsewhere in the boat. So the entire base assembly of each mast was secured by means of "torque arms" to fixed points on the hull, one on the port side and one on starboard. If one thought of the entire *Falcon* rig as a giant energy-transformation device, then the bending and twisting forces exerted on the mast by the wind were ultimately absorbed by the hull. For a sailboat, the engineering hurdle was unprecedented, complicated further by having to fit the mechanisms within the confines of a relatively narrow hull, especially for the fore- and mizzenmasts. Each of the three mast assemblies—the bearings, the torque motor, the various fittings, and the mast itself—weighed close to thirty-five tons.

Perini had already concluded early on that the actual construction of the carbon-fiber rig—the masts, the yards, and the trusses—was too risky for his company to take on. He left that for Perkins to handle, though Perini gave him an entire building at the Turkish shipyard to do it in. But that still left Perini with his worries of making the seventy-five motors for the sails and the twelve more motors for the masts synchronize correctly.

It was all enough to make Perini's head spin, a test of all the hardware and software resourcefulness he had developed in two decades of constructing yachts. "The electrical functions on the *Falcon* would be simple—the machines we manufacture for the paper industry had fifty times the electronic complexity," Perini said. "It is that the simple components become complex in their entirety." On the *Falcon* he'd have to figure out where to place all the wiring and cables for twenty-nine motors per mast, and positioning them to stay out of the way of three masts that were turning. "A vessel can't be designed to be a mast-carrier," Perini remembered thinking back in 2001. "Masts are an auxiliary to the boat—the boat cannot be an auxiliary to the masts." Not until this boat.

———————

There was no way to digitally simulate the winch-and-motor design, with its boltropes and mandrel inside the mast. Putting aside the question of whether a carbon-fiber rig would endure the loads, it was impossible to prove on paper that the electromechanical system would operate smoothly. At the very least, Perkins wanted to see something approximating it in real life. Dijkstra's team assembled a one-sixth scale wooden model of the mast with upper and lower yards. Instead of a motorized winch, they just used a hand-crank to furl and unfurl a mock sail. It took much fiddling, but the system looked promising, though no one could be certain with just a model. The secret of the furling and unfurling system was going to be keeping the precise tension on the boltropes. Too tight, and that could tear the sail; too loose, and that could allow the sail to jam. In a calculated acceptance of risk, the designers assumed this delicate balance could be solved as the project proceeded.

Next, Dijkstra's team jury-rigged a model rig on a fourteen-foot dinghy in the Amsterdam canal where they had previously toyed with a little remote-controlled *Falcon* prototype. To their relief—and to the delight of pedestrians and commuters—this strange-looking vessel, with several people scrunched on board, again suggested that the DynaRig would work. When Dijkstra's team began, the local water patrol wondered what they were up to—this was only months after the September 11 terrorist attacks. But the police officers inspected the gadget and became intrigued by its weirdness. Soon, they were providing a ferry service to escort the *Falcon* model up and down the canal. The primitive sail-control system functioned on different points of sail, as well as when the dinghy tacked and jibed. When the wind picked up, the boat accelerated nicely and stayed on course, and the sail didn't tear. Now all that was left to do was spend four additional years and $130 million to build the rig and vessel to scale—and create Tom Perkins's clipper yacht.

The *Maltese Falcon*, along with *Athena* and *Mirabella V*, would thus be the trio that ushered in sailboats to the club of megayachts. *Mirabella V* was big, *Athena* was bigger, and the one-upping *Maltese Falcon* was the biggest—and strangest. *Mirabella V* reportedly cost Vittoria in the neighborhood of $70 million, give or take some varnish*; *Athena* came in north of $100 million; and the *Falcon*, with her experimental rig that necessitated substantial R&D, cost $130 million. If this wasn't the Golden Age of Sail 2.0, at the very least it was the Era of Excess. While the trio were fairly lumped together as pillars of new engineering and consumption gone wild, their owners viewed the boats largely as rivals. Jim Clark naturally went out of his way to say it wasn't so: that's why the mere mention of the *Falcon*—before you got to "*-tese*"—led him to tell you he wasn't in competition with Perkins's boat, which was the surest sign he was. Joe Vittoria was more transparent, admitting that he'd love to sail side by side with the *Falcon*, though such a contest could never be called a "race." That wasn't due to ego, but because of his insurers, who wrote into his policy that *Mirabella V* wasn't allowed to race.

---

* Vittoria said its real cost was significantly higher and was reflected in its valuation by international brokers at $100 million.

NINE

Dream Boat

If you take the Bosphorus Bridge eastward—from the European part of Istanbul to the Asian side—and head out twenty or so miles into the industrial suburbs, you'll reach the teeming town of Tuzla. This was the center of Turkey's boat-building renaissance—more than thirty shipyards for commercial freighters and tankers, and for superyachts. But in the shadow of humongous cranes against the sky, amid the din of welders and riveters, the old Turkey was still vividly on display: the horses and donkeys and pushcarts competing at rush hour with cars and buses to make it through traffic circles designed by sadists. The locals said the rule was that any vehicle wanting to get into the traffic circle had the right of way, but that any vehicle wanting to *exit* had to wait for an opening. It was a wonder anybody made it out of a traffic circle alive. Perhaps the circles were designed by a French race committee straight from Saint-Tropez.

Two decades ago, Baki Gökbayrak—the head of Perini Navi's operation in Turkey—helped to bring economic progress to this place. Turkey had a rich history of building boats going back to the Ottoman Empire, whose navy solidified its power for centuries. The shores of the Bosphorus were once lined with shipyards making wooden sailboats for commerce and war. It was the advent of steam-powered vessels that helped lead to the decline of the empire. Gökbayrak had grown up in a mountain village ninety miles from Istanbul. His mother was a cleaning lady and his father worked as a hospital orderly. At fourteen, Gökbayrak joined the Turkish navy and stayed for twenty-three years. He graduated second in his class from the country's naval academy; then, at government expense, he spent nearly four years in the United States obtaining prestigious naval architecture and engineering degrees at the University of Michigan and the Naval Postgraduate School in Monterey, California. He still remembered the moment he arrived at JFK Airport in New York City. "I saw those moving stairs and swinging automatic doors—and I decided at that moment that someday I would leave the navy to go out and build such things," he said.

Returning to Turkey with an American bride, and rising to the rank of lieutenant commander, Gökbayrak was a spirited, ambitious assortment of contradictions. Among other things, despite his chosen profession, he hated the water. "I grew up on the land, among fruits and vegetables," he said. "That is where I wanted to stay." His favorite time to be on a submarine or a frigate was before it ever made it into the sea. He viewed himself as a modernist, yet was an avid collector of Qu'rans, including a life-sized edition from the sixteenth century that soldiers carried into battle. He also was a self-professed pessimist who nonetheless had big aspirations to "create something." When he left the navy, he acquired a small Tuzla shipyard that became the Yildiz Gemi facility. Gökbayrak was building two 100-foot motor yachts when Fabio Perini approached him in the late 1980s about bringing Yildiz Gemi into the Perini Navi fold. Together, they then expanded the Turkish shipyard into one of the largest for superyachts in the Mediterranean. It was there that the 289-foot hull that became the *Maltese Falcon* was built on spec.

Gökbayrak recalled meeting Tom Perkins when he first came by to see the hull in the spring of 2001; at the time, Gökbayrak thought little would come of it. When Gökbayrak saw Gerry Dijkstra's sketches for the DynaRig, his pessimism and Americanized sarcasm kicked in. "Sure—this isn't going to happen," he thought to himself. He was not concerned with the size of the project—his workers had done the hull—but rather the new engineering that the rig entailed. Perini didn't have the facilities in Italy to handle years of work on the *Falcon*, so it wasn't an option for him to do anything but the high-end finishing work on the yacht in Italy—work that wound up being done in Turkey anyway. Perkins was obviously delighted to be able to save millions on labor costs, and was comfortable the Turks could handle the work—he had seen the strong Turkish work ethic nearly a half-century earlier when he spent time in the country for Sperry Gyroscope.

In recent years, the labor pool had become more skilled, in part because of the arrival of German immigrants; some naval yards in Germany were in decline due to Asian competition. Gökbayrak's own motivations for the *Falcon* project were born of a minor persecution complex. He wanted his young workers—and the yachting realm—to see that Turkey could take on a historic project. "In Holland, the workers sing songs when they build a boat. Here in Turkey, we just work, we don't smile, we just work very hard. But who could have imagined this here? Ninety percent of my workers don't know how to swim; they've never been in a boat, except maybe a ferry; and they get to eat fish only a few times a year."

The plan at Yildiz Gemi was to get the *Falcon* hull out of her berth at the breakwater and put her in a new, gigantic gray shed to start early work on hull modifications. At the same time, Perini lent Perkins his own shed—what they called "the bone yard"—to create the masts and yards: a complicated process of making molds, cutting and laminating and baking carbon fiber, then wiring and assembling—and hoping. As the final step in Perkins's nearly two years of trying to manage risk in the *Falcon* project, he set out through one last series of tests to ensure that the DynaRig was feasible. He did so with different scale models of one section of the rig: top and bottom yards, with a corresponding mast segment and sail. The models varied in size from one-sixth scale to a full-size replica, and they increased in complexity as well—from wood to carbon fiber, and from hand-cranks to the winches, boltropes and motors that ultimately controlled the final system. Dijkstra's office in Amsterdam already had put together a simplified version of the sail-control system; that wooden model had suggested that the process of furling and unfurling would work. But only a test of a full-scale portion of the rig—with two carbon yards—would prove the sail-control technique for sure. Perkins now was in the carbon-fiber business for real, on his way to spending roughly $15 million on the DynaRig.

The key member of the team was Damon Roberts, a British mechanical engineer who was a pioneer in mast-building, as well as an accomplished racer of sailing dinghies. In particular, he combined expertise in carbon-fiber structures and fiber-optic sensors, the union of which would be essential for the DynaRig. Roberts had been an executive for fifteen years with Carbospars, then a leading European maker of composite masts and other carbon-fiber products (including a pair of stylish eighty-eight-foot flagpoles for the headquarters of MI-6, the British secret service). He was now a founder of Insensys, a Southampton fiber-optics start-up that was aiming to transform the industry of measuring stress in such structures as suspension bridges, the blades of wind turbines, and the long, deep pipes in oil rigs. Insensys did so by embedding "smart" fiber optics in the structures that then yielded real-time data about structural health. Previously, that kind of monitoring could only be done with less reliable strain gauges and other devices. In a mast, the technology worked by integrating .01-inch-diameter fiber-optic cables into carbon-fiber laminate while it was being fabricated. Each thin cable had sensors placed in it. Then, multiple light pulses were sent up the masts through the cables. When the masts were under stress—from bending, twisting, or even compressing—the wavelength of the light pulses was altered, thereby signifying a structural change.

Embedding fiber-optic sensors in a carbon mast—with digital readouts transmitted to the bridge—allowed anyone at the helm of the *Falcon* to know the bending and torsional forces on each mast at any given moment. Dijkstra and Roberts would determine the maximum acceptable loads for both, as well as some combination of the two, and feed those limits into a computer readout at the sailing-control digital panel on the bridge. If the forces on the masts reached 75 percent of the acceptable load (or whatever amount of tolerance Perkins and the designers decided was appropriate), an alarm beeped and the helmsman had to reduce sail area or dump wind from the sails by modifying course or rotating the masts. Roberts analogized the precisely monitored masts to a "real-time, full-size floating wind tunnel."

Instantaneous measurement of stresses on a mast wasn't just about safety. The mast's neural network also gave the helmsman a way to evaluate the boat's performance. Comparing data over time—a computer was able to keep a running track, just as an airplane's flight-data recorder did—would indicate for any set of wind speeds and angles whether the DynaRig was optimally set: Were the correct sails deployed? Were the masts rotated to the best position? The wind on any sailing rig produced both lift and drag, just as water did on the keel. Lift propelled the boat forward. Drag slowed it down. Heeling was good to a point; that gave the boat more buoyancy as more and more of the widest section of the hull was in the water. But past a certain angle of heel, the sails lost power and the distribution of the boat's weight just created drag. By comparing what the mast was doing with how the boat was sailing, Roberts's fiber optics offered guidance on how to sail the *Falcon* faster, far more than the early computer simulations might have.

Such fiber optics would work only on a freestanding mast. On a conventional rig, the multiple points of attachment—the stays and shrouds—made measurements too complicated.

On the *Falcon*, any helmsman still had to push the buttons to trim the masts and pick the correct configuration of sails. Apart from having an old-fashioned autopilot that held a constant compass course, Perkins would never allow computers to control his vessel directly. (There was that time on the maiden voyage when someone on the bridge turned on the windshield wipers—and for some reason the autopilot disengaged, causing the boat to suddenly veer into the wind and send the boat into a frightening twenty-degree heel. But the problem never repeated and, well, it made for some momentary excitement.)

Roberts took several engineers from England to Turkey to get the mast-manufacturing operation underway. They joined thirty-five local workers. Together, they comprised their own DynaRig group, separate from the vast construction project in the big shed at Yildiz Gemi. The hull got the oohs and ahhs—its size guaranteed that—yet the mast-makers faced the more daunting task of creation. Hulls were easy compared to building a rig that nobody had contemplated just two years before. The DynaRig group not only had to make the masts—in a horizontal position obviously—but then it had to figure out how to get the fully assembled pieces out of their shed and inserted into the *Maltese Falcon* hull. As the workforce came together, Gökbayrak remembered thinking that his once-small yard was becoming the talk of Tuzla—as well as an employment hub.

The full-scale test rig—two carbon yards, a mast section, and the sail, along with winches, boltropes and a motor—was a disaster at the start. The rig was mounted on a trestle and set in concrete at the end of a dock where it would pick up more wind. As it did, the sail tore repeatedly. The problems were gravity and friction—two fairly elemental problems facing sailors. The sail got caught on things; it skipped the track; it didn't correctly roll off the mandrel inside the mast; the motors or winches jammed; there never seemed to be the proper combination of tension and slack on the boltropes. And all those problems were taking place on a dock and not in the middle of a thunderstorm or an ocean of thirty-foot seas. "This was the hardest part of the project," Perkins said. This was when we might've pulled the plug, if we didn't get it right."

Ironically, it wasn't the scale of the project—the size of the hull, the height of the masts, the space-age materials or the labor involved—that was threatening to halt production. It wasn't the sail design either. In this campaign to build the most-advanced sailing machine in history, the sails themselves were low-tech: standard Dacron and relatively cheap (about $20,000 apiece). Because the sails were small, because they were supported on top and bottom, and because they never were flailing, they were subject to relatively little load—magnitudes less than the stress on the masts. The only customized features of the sails were their shape and varying weights along the DynaRig. The upper sails were lighter weight, so that if the *Falcon* was ever hit by unexpected heavy winds and the furling mechanisms all catastrophically jammed, those upper sails would just blow out. That way, the boat didn't get knocked down. (In the old days of the clipper ship, the crew had a shotgun at the ready to blast holes in the sails, in case of a squall.) Nobody was especially concerned about losing the *Falcon*: tests in the tunnel showed that even at full sail, the boat wasn't going to be pushed over to the point of no return until a gust hit 122 knots, but the sails were nevertheless constructed for that sacrificial contingency. Doyle Sailmakers, based in Marblehead, Massachusetts—the only major supplier for the boat from the United States, if you didn't count Perkins—made innovative sails for America's Cup boats, many Perinis, and *Mirabella V*; but the *Falcon* just wasn't a high degree of difficulty.

But the low-tech mechanics of getting the sails onto and off the yards was another story. As Dijkstra had discovered with an earlier model, the success of the furling and unfurling system depended on maintaining exactly the right tension on the boltropes, no matter the wind conditions or heel of the boat. When the ropes were too tight, they put undue stress on the clews of the sails, which then ripped. If the ropes were not taut enough, they jammed and allowed the sails to flop around, risking damage to themselves and nearby hardware. Over the course of many months, the designers finally achieved the correct tension.

It helped to have the expertise of Fabio Perini. In the beginning, he had wanted nothing to do with the DynaRig. He just thought its costs and challenges were too open-ended. That's why Perkins was to be responsible for everything to do with the masts and yards. But Perini changed his mind, and took over the sail-control aspects of the rig—the process of furling and unfurling the sails. How could Perini not be involved with the winches and the customized tensioning arm on the yacht that might become the flagship of his company? As it turned out, Perini's tensioning arm—a device using a spring and a switch that regulated tension electro-mechanically—was far simpler than any purely electronic system pursued by Perkins or Dijkstra would've been. After all, Perini was the master of tensioning arms—the inventor who had first perfected his father's paper-feeding machines. Perini needed computer circuitry for the tensioning arm to communicate with the control panel on the bridge, but micro-chips weren't calling the shots. *Stupido è meglio.*

Once the sail controls were functional, the design of the masts them-selves had to be tested in real, three-dimensional space. Roberts built a one-sixth scale version of a carbon mast's lower section, right where it entered the hull. That was the section most at risk, where—because it was unsupported by stays or shrouds—it had to withstand the extraordinary bending and twisting forces that no other sailboat had experienced. On a conventional sailboat, the forces that the wind produced went downward into the boat. With the DynaRig, with no downward compression from stays or shrouds, the forces were sideways. It was something like the risk of having a really big tuna on the end of your fishing pole. As the strength of the wind increased, the forces were cubed. Sometimes, physics was a sailor's friend, as in the case of boat stability improving exponentially as waterline length increased. Yet here, nature was diabolical: as the wind picked up, you had to be that much more careful with the rig. One way Perkins liked to describe the maximum permissible load on the *Falcon's* masts was to imagine eighteen one-hundred-ton locomotives suspended from a rod that was one meter long, a bending force equivalent to that on the big wings of a 747. That was what the masts were theoretically up against in a hurricane-force wind—at least until the boat got knocked down and the wind then blew over the sails rather than into them.

Separate from the bending load, the masts faced major twisting forces in adverse conditions. If there was an unexpected shift in wind direction and too many sails were up, the mast would need tremendous strength to fight this resistance. This resulting torque could reach one hundred meter-tons—not as much as those eighteen locomotives, but still more twist than a mast had ever survived. In short, each of the *Falcon's* three masts needed to be incredibly strong. To measure the strain in the test section of the DynaRig, Roberts had embedded the mast with fiber optics. Like the testing of earlier models, this round also took months—and succeeded in continuing the predictable, and frustrating, process of pushing back the date when the *Falcon* might someday see water.

So did further rounds of sophisticated computer modeling. Computers were indispensable, for they allowed Dijkstra and the other designers to make virtual adjustments to the DynaRig—for example, adding more carbon fiber to the large trusses on each mast, onto which the yards were connected. Once those structural changes were made digitally—without the expense of real construction of prototype after prototype—the designers could run another computer model to see if the rig had become stronger. You couldn't make these changes instantaneously. There was trial and error involved, as the computer could only react to each change rather than propose an all-inclusive solution. Computers could make a hundred decisions simultaneously. But just as they couldn't build something from scratch, they couldn't think ahead by factoring in multiple conflicting outcomes. (That's the main reason weather forecasting beyond a few days is still so unreliable.)

Roberts had brought in a computer consultant who specialized in "finite element analysis." FEA, in use for decades, was the gold standard in marrying engineering with computer diagnosis. It was used for testing such dynamic solid structures as planes, cars, buildings, tires, and even human bones. Using fast-processing supercomputers, FEA divided a structure into millions of tiny mathematical "cells" and then subjected them to forces based on the characteristics of the structure being tested. For example, a car could be subjected to wind resistance, rollovers, and collisions with other vehicles or stationary objects. For the *Falcon*, Dijkstra, Roberts, and the other designers wanted to see the stress points when the masts and yards faced the most bending and twisting forces. Using preset equations based on the material being used, the computer then calculated the forces on each mathematical cell and the material effect on adjoining cells, until the forces radiated through the whole structure.

Once the simulation played out, the computer workstation yielded a moving digital representation of the masts under stress. The color display was based on the different loads: blue was good, red was bad, other colors were in between. Such dazzling arrays were pretty much de rigueur in *Mission Impossible* movies—say, in searching for a weak point in a skyscraper or jetliner at which the villain would aim his missile—not that Tom Cruise ever uttered the words, "finite element analysis." FEA of the DynaRig demonstrated that the weak points were in logical locations: corners, slots and places where anything was attached (like the trusses). Even after more carbon fiber was added to those spots, there was always a relative weak point.

What the designers had to do—and then convince the safety regulators at the Maritime and Coastguard Agency—was to make the structures just strong enough and just stiff enough. It was a judgment call, but there were conventions in the industry—like the 122-knot standard for wind strength. The biggest risk to the masts seemed to be from sailing dead downwind, where a gust of wind wouldn't force the boat to heel more. Under those circumstances, the boat wouldn't dump power. Instead, the masts would just be stressed. For that reason, Perkins decided to design emergency "check stays" that could be temporarily installed from midway on the mast down to the deck. Those stays were made out of rope, which would only be used in heavy air when the boat was heading directly downwind, or in one other highly unusual condition. Sea conditions might result in a frequency of the waves coincidentally matching the resonant frequency of the masts oscillating. There was always some vibration in the mast, because the bottom was set in the hull and the top was moving around; the mast was like a giant vertical spring, with nothing except the sails to dissipate the energy. In that exceptional scenario—which usually would occur only when the boat was under engine power—Perkins would have the crew hook up the check stays to stabilize the masts. Check stays, however, created their own problems, since they prevented the masts from rotating. So if the check stays were in place when the boat was sailing, changing course quickly was impossible.

When FEA simulations were complete, and the required structural changes were made in the one-sixth miniature of a mast's lower section, Dijkstra and Roberts subjected the model to increasingly strong bending and twisting forces. Roberts's fiber-optic readings roughly confirmed the data that the wind tunnel and computer simulations had offered up. Ultimately, the rig was tested to destruction—the most theatrical measure of the mast's strength and the moment of truth that everyone had waited for. "Run all the simulations you want and do the equations," thought Perkins, "but until you saw the thing actually put to the test, you couldn't be sure." Plus, as all tinkerers and amateur scientists knew, it sure was fun to blow things up. With a chain hoist twisting and bending the mast at the top—as it remained fixed in cement at the bottom—the mast finally failed at 15 percent more load than computer modeling had predicted. Nobody at the test had thought much about what happened to carbon fiber when it failed—and that was part of the thrill. Now they would find out. Carbon fiber didn't so much snap or break—it exploded. This was absolutely the last time Perkins or anybody else associated with the *Falcon* wanted to see carbon fiber disintegrate.

TEN

Luxurious Machine

Nearly two years after he had first wondered about the *Falcon* hull, Tom Perkins had now contained most of the technical risk of the project, with an investment of under $10 million. The rest almost seemed routine: modify the hull; add on the superstructure; build the actual sailing rig; write some more checks totaling more than $100 million, and in a few years, he'd have the world's first clipper yacht. It almost worked out that smoothly.

Even if the boat were fast and the rig worked, it wouldn't be much to inhabit if the interior wasn't precisely what Perkins had in mind. He wasn't going to be able to race the boat very often—there wasn't much of anything in the *Falcon's* league—and if he really intended this as his retirement village, it would have to be just so. The research and development for the *Falcon* had been intellectually fascinating but exhausting. The tests in the towing tank, in the wind tunnel, in the canal, and on the ground in Tuzla were often monotonous. Finite element analysis wasn't something that anyone could cogitate over endlessly. Designing the living quarters, though, was the fun part—getting to put lots of frosting on a big cake. The overall look of the superstructure, the layout inside, the high-tech wizardry of the bridge, the furniture, the artwork—these were the things that guests, visitors, media folk, as well as perhaps Joe Vittoria and Jim Clark, would see and remember. This would be the face of the *Falcon*.

Perkins was pleased with how the restoration of his *Atlantide* turned out—the little ship of Dunkirk converted to Art Deco opulence. So he asked Ken Freivokh to take on designing all of the *Maltese Falcon* other than the hull and DynaRig—the superstructure itself and the 11,000-square-foot usable interior. Freivokh, now sixty, was well-suited to the enormous job. He adored big projects (he said he'd always lamented that he couldn't be involved with the ark built by Noah), he understood Perkins, and he shared his hands-on perfectionist personality. His multipaged messages on minutiae were legendary. When the Perini Navi logo was attached to the boat the day after it was launched, Freivokh complained in a much-circulated e-mail that it was "*fifteen millimeters* too low!" He wanted to handle not just the overall look of the yacht and the color of the furniture, but the selection of dinner napkins and the feel of door handles.

Most of all, Freivokh prized originality. When Freivokh walked up and down the docks of the typical boat show and observed the almost identical run of bloated stinkpots—"with the same patio doors, the same deck furniture, the same world of blissful anonymity and conformity"— he thought of the ugly real-estate developments of American suburbia. "I will never do white motorboats," was his mantra. Freivokh was drawn to unconventional wisdom for its own sake, coming up with wonderfully odd ways to try to describe his philosophy. As an example, he'd ask which was better—a milk jug or a teapot. Salespeople concerned only with the basest of values would say it was what sold the most units and produced the highest profit margin, Freivokh said. "I'd suggest that's wrong and the items are designed the wrong way around. The teapot would be a better milk jug, drawing hot milk from the bottom rather than from the top, and the milk jug would make a better teapot, because it served the liquid rather than the tea leaves." Arriving at this conclusion, however, "required a break with preconceptions, which wasn't easy."

Freivokh was a character. He liked motorcycles and toy helicopters (giving Perkins the model he used to buzz Hewlett-Packard chairman Patricia Dunn at a dinner party). He collected designer fish, and punned about things like "being koi." And though his middle name was Lincoln because he shared a birthday with the twelfth U.S. president, he had a bushy handlebar mustache that made him look like a more compact version of William Howard Taft. Freivokh's distinctive voice was mellow and smooth, with an accent more designer-speak than British. And like just about everyone involved with the *Falcon* other than Baki Gökbayrak, he was an experienced sailor, with a particular fondness for the International Dragon twenty-nine-foot day-racers. Freivokh was born in Hollywood, but had become a Brit circuitously. His father was French and his mother was Chilean. They had met in Peru and later emigrated to southern California, but she missed her family in Peru, so they moved back to Peru when Freivokh was a child. He went to school there and worked as an architect, but wanted further technical training. At twenty-four, he won the only South American scholarship for a postgraduate year at the influential Royal College of Art in London. After three months, he transferred to the engineering school, secured another grant for another term, and earned a masters in industrial design in a record two years, rather than the compulsory three. Then he met Liz Windsor, a British national with whom he's been ever since.

With Windsor and a former top Lego builder and several others, Freivokh ran an eight-employee boutique design firm on the south coast of England, not that far from Perkins's Plumpton Place. While Freivokh's much-acclaimed specialty was yachts, he also did private-jet interiors, customized Harley-Davidsons, and one-off projects that included a rustic home for a disabled boy, a spectacular penthouse overlooking Amsterdam, and his own state-of-the-art twenty-thousand-gallon figure-eight koi pond and waterfall. The latter was featured as a cover story in Koi Ponds & Gardens, as "the *ultimate* million dollar pond" (begging the question of just how many others there were). If you wanted to know everything about koi varieties like Hana Shusui, Gin Rin Kigoi or the "inquisitive" thirty-eight-inch Chagoi, Freivokh was your man. While

Perkins had sufficient insight to know Freivokh's obsession with fish was little different from his own with boats, he still liked to tease Freivokh about the pond's "underground control room with more switches and dials than a nuclear sub" and the "fish vet who made house calls." The *Maltese Falcon* may have been the boat that Google (or Genentech) helped to build, but as Perkins said, Freivokh's prized possession was "the pond the *Falcon* built."

———

Perkins learned of Freivokh by seeing photos of his work in glossy yachting magazines. Freivokh knew of his classic schooner *Mariette*, but not Perkins specifically or his financial renown in the United States. When they met at Freivokh's studio in Britain in 1997, Perkins made an instant impression, showing up in his silver McLaren F1, the fastest production car at the time, of which only a hundred were made. This was better than even a Harley, Freivokh thought. Perkins commissioned him to do *Atlantide* and now, a few years later, Freivokh was the only designer he considered for the *Falcon*.

If Perkins knew anything about himself, it was that he hated repeating yesterday's tricks. As much as he loved the Edwardian style of *Mariette*, the Art Deco theme of *Atlantide*, and the traditional design of both *Andromedas,* he had no interest in doing them again. His instructions to Freivokh were to do something "cutting edge, modern and clean." The motif that Perkins had in mind was "luxurious machine." It was his daughter who came up with the notion about Darth Vader's kind of yacht, presumably commenting on the look of the yacht rather than the persona of her father. Said Freivokh: "This was a transportation machine and a technological statement—and it deserved better than to be treated as just another ornate Manhattan apartment or expensive country cottage."

Perkins had a few specifics for Freivokh. He wanted a small superstructure, relative to the length of the boat—far smaller, for example, than on Jim Clark's *Athena*. That in turn dictated more deck space. Perkins additionally wanted no flying bridge atop the superstructure—the outside area from which all other Perini Navi superyachts were typically sailed in daylight and good weather. That meant the *Falcon* could only be sailed from inside the bridge—akin to how motorboats were handled and antithetical to the "wind in your face and sea at your back" glory of sailing. For his part, Chris Gartner, the captain of the boat, worried about being indoors and missing some unquantifiable change in the elements. "You can have all the instruments you want, but it's all on a screen," he said. "If I'm not outside, I can't smell the rain coming, or feel the temperature changing, or hear the waves cresting. That's a risk. If you're a sailor, you want to be outdoors."

Gartner had never forgotten the time on the second *Andromeda*, en route from San Francisco to Hawaii, that they were sailing along on a beautiful night at ten knots when disaster nearly struck. Neither of the crew members on watch was outside, preferring instead the air-conditioned comfort of the bridge. Then, a white squall suddenly rolled in. Like clear-air turbulence in an airplane, a white squall was a freak atmospheric condition. *Andromeda* was hit by a wall of fierce winds that reached fifty knots. The boat was knocked over, at an angle nearing eighty degrees—ninety degrees meant being entirely on her side. There was panic, as *Andromeda* shuddered and the crew had difficulty getting the sails in and down to a manageable amount of sail area. Under those conditions, at a certain point the two masts on the ketch would just snap. "I was certain we were going to lose the rig," Perkins said. Apart from damaging the boat and losing the ability to sail, dismasting could kill people if they were below the rig as it came down. While the crisis passed, Gartner, as well as Perkins, learned a lesson. While a white squall was undetectable on any instrument, they believed that someone outdoors might still have been able to sense a change in the elements. From then on, Perkins's vessels had a rule: when the boat was under way, one crew member always had to stand watch outdoors.

That rule continued on the *Falcon*. But in the end Perkins thought a flying bridge would hurt her sleek lines, would force the elimination of the lowest yard and sail on the mainmast, and would not be an especially comfortable location if his boat went as fast as he intended. Aesthetics, too, played a role in raising the mainmast slightly higher than the fore- and mizzenmasts. To Freivokh's eyes, the rig looked better balanced that way, and the slight variation didn't affect the aerodynamics. The signal mast way in the bow, carrying navigation and communications gear at its top, was Freivokh's design. Given the weight of that equipment so high up, the normal configuration for the signal mast would have been a tri- pod. However, that would've been ugly. Instead, Freivokh devised two "legs" that seemed to wrap themselves around the mast seamlessly.

Finally, despite Perkins's obsession with bigness, he didn't want the *Falcon* to look big from afar. "If you saw my boat on the horizon, I didn't want you to be able to tell the size," he explained. Once you made out a person on deck—as one could see in most close-up photographs of the boat—it was apparent the *Falcon* was immense. A human body gave scale to anything, but without it, perception was tougher. That's why the super- structure's windows seemed to be one continuous piece of smoked glass— they're different from what you're used to seeing on a boat. Similarly, recognizable deck items—like orange man-overboard life rings—are hid- den on the inside of the rail. In Perkins's desire to scale down his boat's appearance, taste seemed to trump ego. A little influence from Freivokh didn't hurt.

For the basic interior design of the *Falcon*, Perkins told Freivokh that he wanted big open spaces and high ceilings on the main level of the boat, where he and guests would eat, talk, read, and spend time. He wanted as few corridors and doors as possible, in part so he could display two dozen large pieces of contemporary art that he already owned—creating essen- tially an urban-industrial loft apartment within the boundaries of a superyacht. And he wanted a little bit of megayacht sacrilege: no varnish. Apart from the cap rail running all the way around the side decks, there would be none on the boat. If ever there was a way to break free of design convention—and spare the crew some maintenance (new varnish was supposed to be applied every six weeks)—this was it.

The rest of the details were up to Freivokh, who came up with a few key principles. He tried to have rooms flow into each other, rather than following the compartmentalized approach that most yachts favored. He used a range of exotic materials, including thirty-six varieties of leather. And he emphasized curves over straight lines: that preference expressed itself most clearly in the gleaming Porsche-silver metallic superstructure, the main contour of which was all curves. Yacht superstructures were typically quite boxy. By contrast, the *Falcon's* lines, as much as its color scheme, are what conjured up *Star Wars.* Freivokh so admired clean lines that he designed extensive louvers to cover up all the unavoidable air vents, access panels, and cowls that dotted the vessel's exterior. Looking merely like simple, decorative horizontal rings around the superstructure, the louvers were there to serve a function. He made no other attempt to hide that the boat was a complicated amalgam of high technologies.

Among the various *Falcon* contractors, Freivokh probably did the best financially. Nobody got especially rich from the entire project—which, ironically, was often the case with superyachts—but he received 4 percent of what was spent on his designs, and that percentage worked out to be about $1.4 million. (The money for designers paled next to what yacht brokers could make in a single phone call—up to 10 percent—if they were selling a boat. But it still wasn't bad.)

Freivokh responded to Perkins's general guidelines with what he always did: a detailed set of drawings that were just a step or two away from actual construction, which would be done primarily by contractors brought in by Perini in consultation with Perkins and Freivokh. Most designers did sketches, and left it up to carpenters, electricians, plumbers, upholsterers, and other outfitters to figure out the rest. Freivokh wanted it all spelled out: shape, thickness, color, material, insulation, and joints. A Freivokh design, based on 3-D computer models into which furniture and modern artwork were placed, was akin to a Steven Spielberg storyboard of a movie: little was left to improvisation or the judgment of those putting the plan into effect. The only wrinkle was that Freivokh didn't always anticipate what it took in man-hours to make a drawing happen in three dimensions.

———————

The most dramatic manifestation of that phenomenon was in the saloon on the main deck. That was the space where guests typically entered the *Falcon*, just off the outside seating and eating deck adjacent to the mizzenmast in the aft portion of the boat. The saloon took up almost half the square footage of the inside area of the main deck. Occupying the full indoor width of the vessel, with a wall of windows on both sides, this was the ideal protected spot for guests. Measuring thirty feet by thirty feet, the room was fully carpeted and had a huge array of burgundy leather couches in a U-formation, facing a fifty-inch plasma screen TV that was hidden behind an imposing silver falcon on the wall. Burgundy was the muted color of choice throughout much of the interior. The couches sported fold-down armrests with built-in cup-holders, though it was generally accepted that if you spilled Coke or popcorn in the saloon, you'd get deck-swabbing duty. On a large glass-top coffee table was a five-foot sculpture of a vintage powder-blue Bugatti by the ceramist Dennis Clive; when the boat was sailing or small children were nearby, the sculpture receded into the enclosed table at the touch of a button. Built-in speakers filled the saloon with music or whatever was playing on the TV. On the aft end of the room was a circular marble-topped bar, with twelve cushioned stainless-steel swivel seats. To one side of the bar was a fully stocked pantry, and to the other was the stupendous $13,950 WMF Bistro deluxe automatic cappuccino and hot-chocolate maker. (Some might say this gadget was more fun than Jim Clark's touch-screen LCD handheld screens on *Athena*.)

There was nothing particularly unusual about the saloon, except that the circular bar had two enormous electronic sliding glass doors bisecting it. In good weather, the curved doors were stored within the walls, and the bar was completely outdoors, though shaded by a ceiling which also served as the floor of the deck above it. This outdoor area of the bar flowed into the large open space centered around the mizzenmast. Guests in the saloon, or seated at the bar, had a view to the outside—to the sea and to the horizon beyond. But in bad weather—and here's where it got interesting—the doors were closed, so that half the bar remained indoors and accessible to guests in the saloon. The other half was outside, exposed to the elements. The doors—each ten feet wide, each weighing a ton— were not only a barrier, but provided privacy. While the glass looked dark from the outside, it was transparent from the inside. It was an ingenious arrangement, taking into account the realities of heat, cold, weather, and sea conditions, but allowing for a vast entertainment area in nice conditions.

Two years, 13,000 man-hours and nearly $500,000 after Freivokh sketched out the bisecting doors, they were completed. "They were just doors—but they changed my life," noted a man who called himself Türkei, the local worker whose job it was to supervise installation. Like most parts of the interior—pretty much everything other than the bridge and the crew quarters, which were handled by local contractors in Turkey—the doors were built by a tiny German company named Sinnex. Located in a village near Stuttgart, Sinnex did high-end work for the likes of the executive boardrooms and elevator lobbies of Goldman Sachs and Time Warner in New York; the Georg Friedrich Händel Concert Hall in the former East Germany and the IS Bank Tower in Istanbul; European express trains; the Jean-Georges Restaurant in Manhattan; and a collection of multimillion-dollar residences around the world.

The saloon doors of the *Maltese Falcon* nearly forced Sinnex out of business. The gag line around the Turkish yard was that "Freivokh" translated into "send into bankruptcy," which was obviously an exaggeration, but underscored the potential gap between blueprint and reality. "I don't know what all the fuss was about," Freivokh quipped when the project was finally completed. "It only took me ten minutes to draw!" Perkins said the problem with Freivokh designs was that they were so attractive. "You look at the drawings and you think they're simple, but they're not." Any miscalculation, of course, wasn't Freivokh's. His drawings were perfectly clear—and transparently complex. Dealing with the innards of a sailboat was always challenging, because space was tight and few lines were straight. Wires, ducts, and pipes presented a routing problem, especially when electrical connections were involved, since there was only so much voltage that could be accommodated in one area. Some planning could be accomplished with 3-D models, but a lot of the work was trial and error. Sinnex presumably underestimated how difficult it would be to execute Freivokh's design. Maybe "Freivokh" meant "sublime," and "Sinnex" translated into "*Ach nein!* We bid too low."

The reason that the doors were an engineering nightmare was remarkably plain. The conventional way to get the sliding doors in and out of the bulkheads was on a double track. Because the doors were so wide, though, they couldn't fit in the wall in one piece—the boat would've had to be twenty feet wider. So each door—the one on the port side and the one on starboard—consisted of two sections that rested next to each other in the wall, each on a separate track. In the obvious system, at the press of a button, motors would direct the two respective sections on each side to come out on their tracks and combine to form a big door, with a slight overlap. But as a statement of aesthetics and elegance, Freivokh wanted the doors to emerge from the wall on a *single* curved track: one section of door fully emerged, and then the next section moved into position and following on the same track. As Sinnex discovered, this wasn't as easy as it sounded. The system had to work mechanically and there were no margins for error, because the doors had to be completely watertight under all conditions, including any angle of heel. In heavy seas, or even if the boat was just heeling a lot, walls of ocean water could crash over the aft area of the deck (the "poop deck"). Without a watertight, pressure-resistant seal at its entrance, the saloon could be awash. It took a lot of water to sink a boat that displaced 1,367 tons, but the insurance and marine-safety regulators had little patience for big doors it viewed as anything less than perfect.

Sinnex completed the project, and the doors worked without a fault, but only after infusions of cash from both Perini Navi and Perkins. It seemed that every other time he operated the doors, Perkins uttered his famous sigh, and thought about how the doors became an interior-décor crisis. The doors were beautiful, their smoked glass bordered by mirror-quality stainless steel—masterworks of engineering for anyone who cared to take a look as the doors opened or closed. But Perkins said, "If I knew back then what I know now, I would have chosen a double-track system in an instant." Freivokh might've been apoplectic.

———————

Just forward of the saloon—amidships—was the centerpiece of the *Falcon*, the place where you felt most like you were inside a futuristic machine. Here was the soaring forty-foot-high atrium—a celebration of air and space—in the middle of which was the freestanding mainmast that extended all the way down deep into the hull. Spiraling around the mast was a breathtakingly unhinged circular staircase, its steps cantilevered from the buffalo-skin walls and unconnected to the mast. The atrium was actually utilitarian: it served to connect the three primary decks. But more vitally, with its thick glass-paneled floors and eight-sectioned skylight at the level of the top deck, the atrium was truly the heart of the vessel. From anywhere in the atrium, you could look up two hundred or more feet to see the five sails on the mainmast. If you had forgotten you were on a sailboat—if you thought that maybe you were just on a water-borne hotel—gazing aloft in the atrium was a reminder otherwise. Of course, if the boat was heeling at fifteen degrees, you probably knew that already, especially if you tried to ascend or descend the stairs without due care. In daytime, natural light flooded the atrium; at night, it glowed from hundreds of tiny electro-luminescent lights positioned in the floors and ceiling.

For most of its length, the mast in the atrium was encased in a highly polished aluminum column. In that way, when the mast rotated hydraulically, you couldn't see the movement, though you surely heard it. The sound was a haunted-house mechanical groaning, much like when the flaps on an airplane went up or down. However, at the uppermost level of the "flying" spiral staircase—just behind the bridge—the mast was exposed. When the mast rotated during an adjustment of the DynaRig, it was both impressive and scary. Any finger that strayed upon the mast right where it entered its encasement risked getting crushed. This was not a good location on the boat to be having a martini.

Architecturally stunning in its own right—a deft integration of vertical space—the atrium contained only three pieces of artwork. On the main deck was an eerie floor-to-ceiling acrylic on canvas by Lorraine Shemesh. The life-size, lifelike painting showed two swimmers underwater in a pool—perhaps not the visage every guest wanted on a boat in the middle of the ocean. But such was Perkins's sense of humor. At the base of the atrium was a menacing metallic shark, an Emmanuel Chapalain sculpture supported by a thin rod attached to a base. Viewed from above, through the glass floors, the great white appeared to be circling. This also was probably not a good location on the boat to be just after drinking a martini. In the corner of the atrium at that bottom level, Perkins placed his own tour de force, an immense cross-section of petrified yew wood with seminal events plotted along its rings: the Norman conquest, the founding of America, the construction of the *Maltese Falcon*. The wood came from a 900-year-old tree that fell at Plumpton Place in a 1987 hurricane. The sculpture—if that's what you wanted to call it—was titled "Time Flies When You're Having Fun," an aphorism he attributed to Einstein. Indeed, when giving tours of the boat, Perkins sometimes tried to convince rubes that Einstein himself had made the sculpture in his later years. Danielle Steel thought the piece was a monstrosity. Evidently, there was no accounting for taste.

On the main deck, forward of the atrium and split on both sides of a bathroom, were a card room and a writing room. Neither was particularly notable and, once the boat was in the water, their value seemed to be as proof the boat was so big that you couldn't use all the rooms. There were chairs and tables in the card and writing rooms, but they were there as much to give guests handholds in case the boat heeled suddenly. On a sailing megayacht, designing a room without furniture risked turning an unlucky passenger into a ricocheting pinball. The most striking feature of both rooms was the art. In the writing room was an oil canvas of Superman's torso by Gordon Smedt, given to Perkins by Steel. (It should've been called "Man of Steel.") It didn't take the DSM-IV to guess why, though Perkins acknowledged that the real superhero had his own pretty cool self-transport system. Elsewhere in the rooms were two paintings by Bo Bartlett. "Lifeboat" portrayed monster waves and a shark about to engulf a boy in a dinghy—making the Shemesh painting in the atrium seem upbeat by comparison. The altogether disturbing "Wonder Tub" showed another adolescent boy in an old European bathtub, staring forlornly at you. Depending on the observer, "Wonder Tub" was homoerotic, unsettling, or just grim. Between it and "Lifeboat," it was a surprise that anyone went through the writing and card rooms at all.

But you had to, if you wanted supper. Just in front of both was the dining room, taking up the entire beam of the boat. Shoji screens and a backlit wine rack adorned the perimeter. The floor was black leather, the walls, creamy goatskin tile. The highlight of the room was a sixteen-foot-long dining table, elegantly supported at a single point by a carbon-fiber and stainless-steel base. The oblong tabletop was made of an aluminum-and-glass honeycomb that Perkins modestly dubbed "Perkinite," perhaps because it sounded better than "Freivokhinium." Perkinite was used all over the boat: hallways, counters, bathrooms.

Above the dining table was another one of those over-the-top design bonuses that nobody at Home Depot ever thought of. Perkins and Freivokh worshipped natural light and here they installed a magically appearing skylight that seemed straight out of James Bond. At the press of a button, six sharp-edged metal pieces simultaneously retracted from the large circular aperture in the ceiling, much as the shutter on a camera lens opened. The iris skylight that then appeared obviously had to be placed somewhere on the deck above. So on that area of open deck space, Freivokh put in a permanent table, in the middle of which was a piece of frosted glass. Guests at that table had no idea they were eating on what was a skylight for the deck below. Any other skylight on any other boat would always be open, so what was the point of this retracting aperture? "Without it," Perkins said, "there was one less thing to show off."

The dining room was the furthest forward area for guests on the main deck. Together, the dining room, the saloon, the atrium, and the card and writing rooms made up the central living areas of the *Falcon*. And it felt pretty much like one big open space, with expansive sightlines—from the aft deck into the saloon and through to the atrium and beyond, as well as up and down the atrium itself. To accomplish that, Freivokh had to tussle for months with inspectors from the Maritime and Coastguard Agency. It was an article of faith at the MCA that compartmentalization—confined spaces, closed staircases, sealed thresholds—were the best protection against fire and water. In return for adding a few more doors, Freivokh convinced the MCA that he couldn't very well add any near the atrium because that area was really part of the saloon. Freivokh would've made a good lawyer.

In front of the dining room, through a heavy sliding door activated by a motion sensor, was a lot of the infrastructure essential to keeping owner and guests happy, but it wasn't prime real estate. There was the upper kitchen and the crew mess (with its very own WMF deluxe cappuccino maker), which led below to the full galley, a crew TV room and the crew living quarters, which were centered around the foremast, which entered the hull here. The captain got his own cabin, but all the others were shared by two crew members: the lower your rank, the smaller your cabin and narrower your double-bunked berth. It had to be done this way, since the boat got narrower as you moved toward the bow. The crew didn't mind their tight spaces so much as the high maintenance that the paneling and fancy fixtures required. Under the crew quarters were the laundry facilities, spacious freezers, and food-storage areas; further down still was access to the great bearings handling the loads on the masts, and finally the bilge. During the fabulous Monaco Boat Show in the late summer of 2006, when Prince Albert came on board for a reception, these areas didn't manage to get included on the tour.

———————

On the level below the main deck—flanking the atrium—were the main sleeping quarters: four staterooms and the owner's cabin. In front of the atrium was the only corridor on the boat. Along that map-lined hallway were doors to the main sleeping quarters. The staterooms offered the same degree of richness and range of materials as the rest of the yacht. The marble trim on the carpeted floors was lovely; the *Falcon*-emblazoned burgundy cashmere throws were ravishing; the horizontal chrome slats framing the flat-screen TV were of a piece with the exterior louvers on the superstructure; the Italian-walnut paneling, with Macassar ebony highlights, made the rooms airy, and leather ceilings helped deaden any vibration.

The bathrooms were even better—a combination of limestone, brushed aluminum, and two kinds of marble. Neither Leona Helmsley nor Marie Antoinette would've had any complaints. The shower stall was lined with black carbon fiber—as was the assembly for the toilet-paper dispensers. Freivokh was passionate about the head. And since just about any guest on the *Falcon* was bound to get seasick some time—other than Perkins, of course—the bathroom might as well be a nice place to recover. One afternoon, I kiddingly asked Perkins why his boat didn't have those status-symbol $5,000 Toto toilets (with a sixteen-bit processor and 512 kilobytes of RAM under the hood for raising the seat by wireless remote). A James Bond lighting fixture, but no royal flusher? He didn't laugh, and contacted Freivokh to ask why not. Freivokh thought the devices were excessive. For any future meddling on my part, I was threatened with the loss of lavatory privileges.

The owner's cabin, aft of the atrium, beat them all. It extended from one side of the boat to the other, making it as large as the four staterooms combined. It had much the same décor as the staterooms, except more. There was a king-sized bed, a long desk, leather armchairs, and two bathrooms—one had a cavernous circular shower dispensing rainwater-like streams, the other had a spa bath. Through a sliding Perkinite door was the exercise room, equipped with a treadmill, massage table, and two steam cabinets (the kind you sit in, with your head sticking out like a Pez dispenser). And there was art galore—just as creepy as elsewhere on the boat. Perkins was a fan of the Danish artist Kurt Trampedach; two of his mysterious self-portraits faced the bed. It could've been worse. Perkins had owned a Trampedach self-portrait, "Now I Lay Me Down to Sleep," which depicted different images from a child's nightmare. At a house Perkins used to own in Bermuda, his maid was so spooked by it that she refused to enter the room where it hung.

Behind the owner's suite were two small cabins providing four additional berths—for a bodyguard, nanny, nurse, extra guests, or unwanted cousin. And behind that was the voluminous engineering suite, containing the two engines, generators, water-desalination plant, air-conditioning equipment, and the lazarette of water toys. This was also where the mizzenmast entered the hull.

All that was left on the boat, apart from open deck space outside, was the uppermost and smallest level. There were but two rooms there, on either side of the atrium's highest level: the VIP cabin and the bridge. The VIP cabin was behind the atrium. It was smaller than the owner's cabin below, but in many ways it was preferable. Since it was way above the waterline, it received far more natural light, and the back opened up through wide glazed doors to a private railed sun deck, with lounge chairs and a panorama of the sea. And because it was just steps away from the bridge, it allowed Perkins quick access if there was a problem at night while he was sleeping. In fact, he tended to use the VIP cabin especially during voyages, and it was usually referred to as the "passage" cabin. Having already written a saucy novel in 2005, "Sex and the Single Zillionaire," it was in this cabin, Perkins snarkily commented on a *Falcon* blog, that he might pen a more serious work. "Perhaps I will," he wrote. "The plot will be about the struggles of the poor." No room in which Perkins spent a lot of time was complete without weird art, and the passage cabin had something called "Witness: Women and the Gun," by painter Dick Lethem. It was colorful—and violent.

In front of the atrium was the bridge. If the atrium was the heart of the *Maltese Falcon*, the bridge was her brains. This was where you controlled the DynaRig and steered the boat, whether using the sails or the engines. There were screens galore to measure sailing performance and to monitor stresses on the mast, but if you wanted to see any part of the rig, except the foremast visible through the windows up front, you had to go outside for a look. Most oddly, you steered not by some traditionally spoked varnished wooden wheel—which Perkins viewed as anachronistic on a postmodern yacht—but with a small simple dial, about the size of the knob you used to turn on the heat in your car. You had no sense of the rudder, no sense of how a course alteration affected the feel of the boat— in short, you didn't quite feel like you were sailing. The skipper could just as well be playing Pac-Man in a video arcade. That was the price to be paid for a sailing megayacht built to go fast. Calling this command center a bridge seemed to make perfect sense because its layout called to mind the old *Star Trek* bridge. (Freivokh's 2006 holiday card showed the *Falcon* bridge with superimposed digital images of Kirk, Spock, Scotty, and Dr. McCoy.)

Looking forward out to the sea, there was a wall of seven sloping windows extending from one side of the bridge to the other. Below the windows were five large screens for navigation: multiple radar systems, a water-depth indicator, a digital nautical chart that used GPS signals, and thermal-imaging night-vision for the view ahead (getting the last gadget required a prospective owner to meet certain security clearances). These screens were needed for any kind of sophisticated navigation, whether under power or sail. Right behind this array of screens were the engine controls and a steering dial, housed in a "hemisphere" of stainless steel that was a work of art. In reality it was made up of dozens of pieces of stainless steel welded together and then polished to produce a mirror finish, but the hemisphere appeared to be seamless. It wasn't the bridge's equivalent of the Sinnex bisecting doors in the saloon, but it was a labor-intensive decorative touch that turned into many more man-hours than anyone expected. There were two big racing-style chairs on both sides of the bridge. (High tech has its limits. While the chairs had all manner of adjustments, it took the crew a month to discover that a bar underneath could prevent them from spinning like tops while the boat was moving.) The floor was African wenge wood, with contrasting aluminum strips between each plank—it was the only indoor wooden floor on the boat.

Behind the hemisphere and engine controls, in the center of the bridge, was a large island. This was the control station for the DynaRig, with its own high-backed chair. This was Perkins's seat of power—what he had spent tens of millions on and waited years for. Set off by a heavily padded black-leather raised horseshoe, the sailing-control station—about the size of a office desk—was conveniently angled toward Perkins (or whomever was at the helm). It consisted of four large LED screens, as well as a series of digital readouts and the bridge's second steering dial. There were also the critical buttons for furling and unfurling the sails, and for rotating the masts.

The most important of the LED screens displayed a digital representation of the entire rig: the three masts, the eighteen yards, and the sails that were actually in use on the rig. If you wanted to change the sail set, you touched the relevant sails on the screen, which made the selected sails light up. Then you pressed a green button to execute the change. The screen graphics matched what the sails were doing—furling, unfurling, or perhaps getting stuck along the way. A different screen diagram showed the angles of the yards relative to the centerline of the hull, along with their position in relation to the wind. To rotate the masts and the yards, you had to press a different set of buttons. And because the yards could jut out well beyond the width of the boat when the sails were square to the wind—a big problem if another boat was nearby—a safety feature was added to prevent accidentally rotating the masts. Like in the old nuclear missile silos of the Cold War, a special key had to be turned before the system was operational. With these two screens, and the dial controlling the rudder, Perkins or anyone else could single-handedly sail the boat. The process was automated and electronic, but a human was in control. Computers only made recommendations, and it was that distinction from Clark's *Athena* that Perkins insisted on.

He also insisted, against the suggestion of Perini Navi, that the two most important screens—controlling the sails and the yards—remain constant. But the other two screens in the sailing-control station had variable graphical displays, ranging from suggested sail configurations for a given course and wind speed, to the performance of the boat against the theoretical optimums indicated by wind-tunnel testing, to the tension on the boltropes that were used to deploy each sail. Most often, one of those screens had on it the pride and joy of the Insensys smart fiber-optics system. On a rising bar chart that appeared on an image of each mast, the screen showed the bending and torsional forces being exerted in real time. That data was combined into an overall load figure, which was then measured as a percentage of the maximum stress permitted on the masts. If the forces reached 75 percent, the bar chart on the screen flashed in red and an alarm went off.

In addition to the four screens, digital readouts gave the wind speed and direction, as well as its angle relative to the bow of the boat; the heel of the boat; and, of course, the *Falcon*'s speed. If you couldn't be outside steering the boat, the sailing-control station was the next best thing. And just in case anyone wanted to work the DynaRig from outside, Perkins built a portable, wireless control-panel that could operate sail deployment and mast rotation. This battery-powered device was especially helpful for a crew member ascending the rig to make repairs.

The funny thing about the *Falcon*'s high-tech bridge was that neither Perkins nor any of the crew called it that, despite its resemblance to the *Star Trek* command center. Instead, it was the "wheelhouse"—despite the absence of a wheel. The reference might've been part affectation, but primarily it was Perkins's tribute to the clipper ships.

———————

By the spring of 2006, more than five years after first seeing the model of the 289-foot hull that Fabio Perini had in his office, the *Maltese Falcon* was nearing completion. More than 300 workers and craftspeople had put in 1.2 million man-hours. The mast-builders alone had logged roughly 200,000 man-hours. In the world of superyachts, the calendar was routine. Everything took longer than it was supposed to, and everything cost more than anybody anticipated. It was more like building a small city than a house—with its own electric plant, water department, fire department, communications center, and food stores (including one 216-cubic-feet walk-in freezer) that had to be self-sufficient in the middle of an ocean.

But with Perkins convinced by the second year of testing that the DynaRig was clearly possible, the remainder of the construction project was within the realm of convention. The mundane process of painting was laborious, but it was to be expected: Perkins wanted gloss-black—the hardest of colors—and he wanted the finish to be flawless. Dark colors tended to reveal every imperfection, which is why most yachts were white. The actual paint job was easy: a coat of spray-paint and then clear-coat. But preparing the surface by "fairing" was not. For two months, workers added ten tons of epoxy filler to even out little bumps in the hull and make it as smooth as an eggshell. Then came two solid years of long-boarding: a low-tech process that used ten-foot-long, curved pieces of wood with handles—and Turkish workers—on one side and sandpaper on the other. A long-board, particularly on a large, curved surface, was more likely to pick up imperfections that the eye could not see.

The only hiccup that threatened to shut the project down was financing—one of the things Perkins knew best. As it turned out, he had significantly underestimated the fall of the U.S. dollar relative to the euro, which was the currency that he was paying Perini, Dykstra, Roberts, Sinnex, and just about everybody else. Perkins had hedged the dollar-euro equation, but not enough—and it wound up adding about $15 million to his costs. He was willing to absorb that, but early on he decided that he didn't want to have to liquidate any of his stock holdings, including his position in a promising young company called Google, to make up the shortfall. Instead, he wanted Perini Navi to finance some if it by altering how much cash flow it required from him. At one point, for a two-week period, he actually canceled the project, before Perini Navi accountants agreed to more favorable terms and everybody was reasonably content. Perini Navi surmised—likely, correctly so—that even in an emotional matter like a boat, a cold-hearted venture capitalist might really walk away. In the end, the ascent of the euro had an upside. Unlike with his other yachts, Perkins had decided to charter the *Falcon* to guests for ten weeks or so a year. It was a way to help defray the millions of dollars in annual overhead for the boat—crew, insurance, and regular mainte-nance—as well as to give the crew a chance to reap a lot of extra income. The *Falcon*'s weekly charter fee was €335,000, which by early 2007 trans-lated into $440,000. (Charter guests paid €335,000 *plus* the cost of fuel, food, liquor, dockage fees, and roughly a 10 percent tip to the crew, pro-ducing a total bill of up to $1 million.) For ten weeks a year, that was a lot of Perkins's expenses defrayed.

ELEVEN

Yachties

*T*alitha G was the quintessence of the classic motor yacht. Launched at the end of the Roaring Twenties, she was the 262-foot steam-powered water toy of various scions, before winding up in the hands of billionaire J. Paul Getty Jr. in the late 1980s, who named the boat after his tragic, late second wife. With her two wide funnels, decorative bowsprit, and stately lines, *Talitha G* was easy to spot in any harbor of the world. Tom Perkins and the second *Andromeda la Dea* found themselves parked right next to her one spring weekend in Antigua a decade after Getty got her. Perkins's own guests had gone home, so he was there alone with the crew. On the other side of *Talitha G* was another superyacht, which was leaving for Germany the next day. A major ocean voyage was always cause for a party—not that one needed an excuse—so *Talitha G's* crew sent the boat off with air horns, champagne and beer, and a water cannon. The only problem was that a lot of water got sprayed on *Andromeda*, whose deck had just been varnished. *Andromeda's* crew wasn't happy and wanted to exact a little revenge—not that one needed a reason for some high jinks, all the more so when a big stinkpot was involved, even if there weren't any Gettys aboard.

A few days later, *Talitha G* herself was getting ready to set off across the Atlantic. The *Andromeda* crew couldn't think of much beyond a smoke bomb or a water balloon. "What about a greased pig?" suggested a voice below deck. It was Perkins. "It's not up to me, but if I were consulting, what I'd really recommend is a greased pig."

"What?" asked an incredulous engineer.

"No, really, it would be pretty funny," Perkins said. "A greased pig would make a big mess and they'd never catch it." He was right, of course, and a pig carried with it none of the superstitious baggage of a bunny rabbit.

But where does one find a greased pig on a Caribbean island? You asked one of the locals, named Kong, a guy who lived in a shack he called the Empire State Building. Kong catered to the superyacht crowd down by the water and was known as one of the best refinishers in the world. Wealthy owners sometimes flew him to Europe just to work on a hull or deck. In Antigua, he was a jack of all trades, adept at procuring whatever a crew needed or doing the odd job that no one else wanted to. It so happened that Kong's mother had an entire litter of piglets. So the crew arranged to rent a pig—the fee was a few drinks—which they brought aboard *Andromeda* under cover of darkness the evening before *Talitha G* was supposed to depart.

Andromeda was lucky to have a deckhand named Dave, who had grown up on a farm in New Zealand. Dave was a pig expert. He knew just how to pick up a pig—by one ear and its hind legs—which wasn't something you could say was in the skill set of most crew on superyachts. In that upside-down position, for reasons only it knew, the pig actually appeared content. That alone amused the *Andromeda* crew, though lots of beer by that point probably helped. Perkins wasn't going to let some pig have the run of his ketch, so the crew put the pig in the closet where life jackets and other equipment was stored. All they needed now was grease, which Kong helped them secure. It had to be water-soluble grease, since if you were going to interfere with the Getty boat's departure for a four-thousand-mile passage, you had to be merciful and only use grease that might take a day or two to clean up.

At two in the morning, the crew retrieved the pig and slathered him up on the dock. By all accounts, he was still having a ball. They put the little porker on *Talitha G* and ran. Perkins had decided that maybe his role as instigator was ill-advised, so he remained below in *Andromeda*, watching furtively through a window. Like some superyachts owned by recluses or movie stars, *Talitha G* had infrared motion-detectors on deck. It would be no time before an alarm went off and the crew came running. But for fifteen minutes, the pig ran wild up and down the run of the 262-foot yacht. The alarm stayed silent. So much for the idea of a vigilant night-watch on the boat. *Andromeda*'s crew thought this was hysterical. Perkins howled. The only thing that would've made the scene more perfect was if *Talitha G*'s most prominent charterer—Tom Cruise, with whichever girlfriend—had been aboard.

What to do next, with the pig on the loose and the crew tidily tucked in below? Two of the crew members from *Andromeda* went back to *Talitha G*, climbed aboard, and banged on as many metal objects as they could find, then ran off again. The alarm sounded and *Talitha G*'s groggy crew came out to find chaos. Four big guys tried to catch the pig, which by this point was having about the best night of its life. The pig evaded any attempt at capture. The crew went to get reinforcements. The pig kept running, spreading grease everywhere. The crew gave up and went back to sleep—no small wonder, given the squeals resounding from their deck. At sunrise, the *Talitha G* crew came out in full force—twenty of them—with stacks of towels, gloves, and a fishing net that they successfully used to corner the pig half an hour later. By this point, the crew of many of the boats on the dock had heard the show and came by to watch. So did the crew of *Andromeda*. Nobody suspected them. Not even when Dave the pig-wrangler wandered over, trying to innocently notice the commotion while also knowing that the rental period for Kong's pig was expiring. "Oh, you have a pig!" he told the *Talitha G* crew. Dave obligingly demonstrated how to carry a pig—all the way to a waiting taxi back to the piglet farm.

It gradually dawned on the *Talitha G* crew that Dave's smooth entrance and exit wasn't coincidental, and they vowed revenge. *Talitha*

G's captain confronted his counterpart on *Andromeda*, demanding to know, "Now why did you do that?" To which the *Andromeda* captain replied: "You ruined our varnish!" When *Talitha G* finally left Antigua, still not completely clean, it launched a water assault on *Andromeda*, which by this time no longer had to fret over wet varnish. As *Talitha G* headed east, *Andromeda*'s tender followed in hot pursuit with a carton of eggs. The game was to land an egg down either of the smokestacks on *Talitha G*. Scrambled or poached for everyone! Days later, with *Talitha G* well into her transoceanic voyage, *Andromeda* suggested a truce. *Talitha G* accepted. The *Andromeda* crew that ended up on the *Maltese Falcon* still worry if the peace treaty is permanent. To this day, Perkins remains convinced that none of the Gettys ever suspected that *Andromeda*'s owner was the mastermind of the great greased-pig caper.

No Gettys could be reached for comment for this book. Neither could the pig.

As owners went, Perkins was just about the ideal boss. He knew the crew, he talked with them, he ate in their mess when he didn't have guests. From his perspective, a satisfied, happy crew was the ultimate luxury—offering a kind of attention that not even five-star hotels provided. From the perspective of the crew, life on a superyacht was the occasional pig caper in an exotic locale mixed into a routine of food service and cleanup, manual labor on deck, engineering work below, and cramped living quarters up toward the bow. Within that subculture of self-described "yachties," the sixteen men and women who staffed the *Maltese Falcon* were a varied bunch. They were mostly in their twenties, though one was over forty. They came from the usual English-speaking countries like Britain and New Zealand, yet the captain was an American who'd been with Perkins for sixteen years and whom Perkins regarded as a second son, albeit a son he employed. Some had college training, but most had headed off after high school, eager to take the road not taken by their conformist classmates.

The crew members were mostly sailors, although some had worked on stinkpots, including Paul Allen's motor yachts, before deciding they preferred the aesthetics of sailing. There were the usual array of engineers and stewards, but there also was a "spar master," whose primary job was caring for the DynaRig and climbing up it when things got stuck, as well as a massage therapist for Perkins's chronically sore back. One crew member from Malta had been shipwrecked after his sixteen-foot catamaran capsized, and he spent four days adrift with his girlfriend before being rescued by an oil tanker. Another was a barmaid and chef at a pub on the coast of England who was whisked off to sea, where during a world crossing she was held captive by South American pirates who commandeered her owner's yacht. Another was born on a boat, served as a roadie for an Irish rock band, and then helped build weapons systems for the British government. And another—Christian Truter, the young Australian-born spar master who grew up in Singapore—was an extra in *Master and Commander* and so much the sculpted stud that the star Russell Crowe urged him to forgo a sailing career in favor of the movies. For the moment, Truter said he liked nothing more than being able to climb the masts and observe the sea and the *Falcon* from two hundred feet aloft. "Climb a mountain, climb a tree, climb the rig—it's quiet and it's peaceful," Truter said, though he acknowledged that being near the top of the rig in a fifty-knot gale, with the mast bending up to ten feet in any direction, was not likely to be so serene.

Most of the crew looked the part of sailor: rugged, tan, attractive, tattooed, with just the right amount of streaks in their blond hair. And they had the usual assortment of habits of almost all yachties. "World-Class Docking!" tittered an ad in *The Crew Report*, a magazine that catered to crew lifestyle. Yachties loved the bar scene in port; they loved to hook up with each other in what often seemed like a relationships merry-go-round; they loved to gossip about who got drunker and hooked up more. Indeed, they liked to say the fastest communications network on the seas was not the Internet or the radio, but the chatter between yacht crews. Mostly, though, they shared an affection for warm-weather adventure and the conviction that the sea would give them the adrenaline rush of their lives. As crew on the *Falcon*, they were royalty. Between crew agencies and word-of-mouth e-mails, more than two hundred resumes had come in when the last positions on the boat were filled just prior to her launch in the spring of 2006. This was the hottest job in megayacht sailing. When the crew walked into a seaside bar, they almost always wore a *Falcon*-emblazoned T-shirt or fleece jacket. It was the surest way to get noticed.

But those lives—be it on the *Falcon* or another superyacht—reflected a paradox. While it took considerable courage to pick up and leave family and friends for a faraway existence, it also meant buying into a lifestyle that gave them little control over that existence. In pursuit of apparent freedom, they had to resign themselves to not having any. To the extent they were self-aware enough to recognize the contradiction, most of them reconciled it by embracing the incredible structure that the lifestyle gave them, with little real responsibility. A crew member had no apartment or house to take care of, no meals to cook, no bills to pay, and few plans to agonize over. It was the atypical member of the crew who had a spouse back home, and the rarer one with a family. Room and board were free. So were most of the clothes. Health insurance was provided. You didn't have to shop for furniture or rugs. You didn't need a car. About your only major expense was the bar tab—that's why it was said of "grotty yachties" that the trouble with blood was that it interfered with their alcoholic system. If you moderated your indulgences, the tab might only be several thousand dollars a year, with plenty left over for next year if by then you chose not to moderate.

Best of all, by most crew members' way of thinking, somebody else set the "program," as crews called the itinerary of a yacht. Because of dockage reservations and immigration regulations, the program usually was set well in advance, but nothing stopped an owner from picking up anchor and setting sail on a whim (something Perkins did more than the typical owner); that was part of the point of having a superyacht. Your only job was to do what you were told when you were told to do it, in return for getting to do it while setting sail to some of the world's great out-of-the-way destinations. It was true enough that you might get little time to see the sights—which normally meant exploring no further than the bars that lined the harbor, since on most days you were busy on the yacht with scrubbing, polishing, folding, and fixing—but nonetheless it provided a regimen.

On top of all that, the pay wasn't bad if you could move up beyond deckhand—"chamois technicians," as some called the men who swabbed the decks—or stewardesses, the women who made the beds, cleaned the bathrooms, and served the gourmet meals. Most captains—usually men—made more than $100,000, and a few made close to twice that. First mates and engineers weren't far behind. Chefs might make more than all of them, if they were highly sought after. None of that included tips from charter guests, which could add up to tens of thousands of dollars a year for even a deckhand. And depending on your citizenship, much of the money was exempt from income taxes.

There was an interesting theory among some crew agents that many of the men who worked on yachts were talented dyslexics—people who failed at traditional schooling, but who were good with their hands and quick-witted. Few were bookworms and fewer still seemed to have larger intellectual interests beyond their own orbits. Perkins recalled one young man, with whom he was sharing night watch on the bridge during an ocean passage. Out of the blue he asked, "Tom, what's the difference between the theory of special relativity and the theory of general relativity?" While hours of night watch could be monotonous, it rarely led to disquisitions on Einstein. In his normal explanatory mode—part professorial in that he could assist, part frustrated that you didn't know—Perkins answered the relativity question in great depth, after which the crew member noted, "Yes, I thought so."

Most crew members talked of a life beyond the *Falcon* someday. Natalia Singleton, the ebullient thirty-year-old assistant chef and resident onboard flirt, planned to eventually open a bed-and-breakfast in New Zealand or become one of London's great chefs. Katrina Arens, the thirty-two-year-old British chief stewardess, vowed to have a family someday and not become one of the embittered women who complained at forty that they'd sacrificed too much by going off to sea. "The industry is full of very sad women who never grew up—like a female Peter Pan," she said. Arens worried that a future landlubbing employer might regard her time at sea not as evidence of independence, but frivolity. Articulate and reflective about what she was doing, Arens did not want to be viewed as the nautical equivalent of a ski bum—she worked killer hours, especially when there were guests on board. Her boyfriend from the Canadian Rockies was Justin Christou—captain of *Atlantide*, which Perkins still owned and typically traveled with the *Falcon*—which worked out nicely, as long as they remained a couple. Arens and Christou had met while working on Jim Clark's *Hyperion*. Even though the crews of all super-yachts numbered into the thousands—as the number of big boats dramatically rose in recent years—it was an incestuous labor pool.

Still, by early 2007, Arens and Christou were through. She had left the *Falcon* to work on land—and had run off with the boat's cook as well. (One might say the dish ran away with the spoon.) The chief engineer was gone, too; forty-four-year-old Jed White was the dean of the crew and left to finish building his dream house in Ireland, play more golf, rejoin a life with his wife, and start a family. White had been with the *Falcon* project for five years, dating to early testing of the rig in Turkey, where he lived during the yacht's construction. But Arens and White were the exceptions. Getting out of the yachtie business was hard, precisely for the reason that many yachties got into it: life onboard provided order and the prospect of life on the mainland most assuredly did not. It was hard to take the plunge.

The captain of the *Falcon* was a smart and colorful forty-two-year-old party boy from the mountains of Colorado by way of Newport, Rhode Island. His name was Chris Gartner. He embodied much of the paradox of being a crew member, though his route to the *Falcon* reflected a preternatural degree of drive and savvy. He was the kind of nautical colleague who could do a cannonball jump into the water to douse the decked-out chef on the gangplank on her way to a date on shore—and the next day, in front of the crew at mealtime, accept with equanimity a lemon meringue pie in the face as retribution. But he also knew how to manage the crew's relationship with an occasionally mercurial owner, and when that relationship most needed managing. This was not an *Upstairs, Downstairs* yacht, like so many motorboats with their hidden corridors in the bowels of the vessel, so that crew was seldom seen or heard by the owner and guests, except when serving canapés. "There are some yachts where you're not treated as a human being," Gartner said. "You're a delivery system for martinis."

Perkins's crew didn't have to wear white dress uniforms with silly *Love Boat* epaulets—black *Falcon* polo shirts and shorts were fine. They didn't have to call the owner "Sir"—or, as was the case on one yacht behind the owner's back, "Mr. Asshole." Nor was the owner a weirdo, like the heiress on a well-known yacht who required that when she took a swim her retinue get in the water with her to form an escort to ward off jellyfish. Or like the man who had to have Persian lemons from Harrods of Knightsbridge specially flown in; or the owner who carried separate wardrobes for his wife and for his mistress, and had the crew do the complete exchange depending on who was coming aboard for the weekend; or the guy who had a customized "ionizer" that had to be used on all his food; or the man who wanted his yacht's water tanks filled with Evian, insisting that the crew do so bottle by bottle. Arens said being part of the *Falcon* crew was like being an au pair—serving somebody else "without being a servant." The au pair reference might've been a telling one, if she meant that Perkins was just another great big adolescent to look after, but that probably wasn't what she meant.

To be sure, the *Falcon* was hardly an egalitarian regime. There was the owner and there was everybody else. When he had to wait for his afternoon martini too long, he could be surly about it. He had a cabin larger than that of all the shared crew quarters combined. And the crew, of course, chattered about him and his guests. But relatively speaking, most of the time there just wasn't that much intrigue on board to interfere with the smooth functioning of the boat. Some of the credit for that went to

Gartner. All it took to be an owner was money. To be a captain required more. Handling crew politics and psychodynamics—within the crew and in its interaction with Perkins—was as important to the functioning of the *Falcon* as the DynaRig. "Chris is the lifeblood of the boat," said White, the chief engineer. "He modulates Tom."

In the summer and fall of 1991, Gartner was living in Newport, Rhode Island, a hub of sailing in the United States. It was a logical place for him to be. Several years earlier, he had graduated from the University of Colorado at Boulder—with a degree in public relations and absolutely no interest in finding a job in that field. He had long been a sportsman, competing on the freestyle-mogul skiing circuit and reaching the rank of eleventh in the world before blowing out his knee. He had been a sailor as well, spending winters as a kid in Fort Lauderdale on a twenty-nine-foot boat his father had built. The father was a surveyor, for whom there was little work after the Colorado snows arrived; so he brought the boat and the family to south Florida, and they spent six weeks cruising the Caribbean. Gartner took some correspondence courses to avoid being declared a delinquent and then returned with the family to Colorado for school and ski camp.

The end of his skiing career led him back to boats. He spent most of the year working at ski resorts in the Rockies, then in summer headed for the East Coast for boating jobs that ranged from running a forty-foot sloop to doing odd jobs in boatyards, like sanding hulls and painting bilges. On one trip aboard a Hinckley 42 to the Caribbean, for which he got five hundred dollars, he thought to himself, "You actually get paid for this?" As the 1991 season was ending, he saw Perkins's second *Andromeda la Dea* pull into Newport, fresh out of her wrapper as she completed her first voyage across the Atlantic just in front of the "perfect storm." Gartner knew she was one of the largest sailboats in the world at the time and her arrival was the talk of Newport. She'd be there for weeks of outfitting and repair, and then would continue on toward Cape Horn and her trip around the world. "As you walked along Thames Street, all you could see were those two silver masts towering over the rest of the harbor, that beautiful dark-blue hull," Gartner recalled. "It was the most incredible sight I had ever seen. I just wanted to work on her so badly."

The next day, he walked down to Bannister's Wharf and found *Andromeda*'s captain, Don Lessels, a bearded Scotsman who was having his afternoon tea. Gartner asked the captain for any kind of day work. In fact, Gartner asked every day for more than a week. Lessels didn't have a lot that needed doing and if he did need someone, it wasn't likely to be all five feet and seven inches of Chris Gartner. On the day before he was leaving for a winter job back at Vail, Gartner went by *Andromeda* one last time. Lessels and his wife—the chief stewardess—finally agreed to interview him the next morning, hours before his flight home. They said they might have a trial position for him as a steward. "Do you have any silver-service experience?" he was asked.

He once waited tables at a restaurant, so he figured that almost counted. "Yes I do!" he said. "But I'd rather be a deckie." While the captain found Gartner's relentlessness oddly charming, he didn't offer him a job. Gartner returned to Colorado and his girlfriend Heather there. Two weeks later, the phone rang at their house. It was Lessels, asking if he could be in Newport in a few days. The job was what Gartner proposed—working on deck and below. Gartner obviously replied that he'd be there. "Containing risk" wasn't part of his vocabulary. Here he was giving up the life he knew, and burning professional and personal bridges. His father, vicariously enjoying the wanderlust, urged him on: "You'll see the world, you'll save some money, you'll get to sail." Gartner told Heather he'd be back in six months, which wasn't quite correct, since if the job worked out, he wouldn't be back for a few years, if at all. That was Gartner's rascally way with women, which became legendary in yachtie tittle-tattle. When a Perkins vessel came into port and the barmy beers came out at the local pub, the crew of other boats asked the usual questions: Where's your boat been? Where are you going? Who's in love with whom? And in this instance, who was the latest in the trail of tears that Chris Gartner left across the seven seas?

Gartner showed up on *Andromeda*, Heather moved on with one of

Gartner's best friends, and that was that. When the boat reached the Caribbean a few weeks later, Gartner got to meet Perkins, who came aboard with Gerd and a group of friends. Given that Perkins owned this multimillion-dollar superyacht, apart from being a Silicon Valley kingpin, Gartner was terrified. There were no employment contracts at sea: if the owner didn't like him, Gartner worried, he could be out of work overnight. But unlike many other yacht owners, Perkins made a point of talking to new crew members and trying to be normal. He asked Gartner if he liked to sail, Gartner said he did, and Perkins told him he'd work out fine. When Gartner once left a gouge in the side of the boat by mishandling the hydraulic swimming ladder—to another kind of owner, scratching the hull could be a capital offense—Perkins thanked him for owning up to the damage and simply told him to fix it.

While Gartner pretty much knew his way around a sailing ship, he had little idea how to behave while serving meals. He still didn't know what silver service was and, as he found out, he lacked refinement in the dining room. It was nerve-racking. Early on, Gerd had a group of friends from San Francisco aboard who enjoyed their Zinfandel. One guest had drained his glass yet again—by Gartner's count, the fourth time—and Gartner came by with the wine bottle, wrapped nicely in linen, and asked, "Would you like *another* drink?" Gerd was mortified. She whispered to Gartner a few minutes later, "Even if the guy's completely drunk, always ask, "Would you like a *fresh* drink?" Regrettably, Gerd's voice wasn't soft enough. The whole table heard the discussion and broke out laughing. The guy had five more drinks.

His future clearly not as a galley slave, Gartner progressed from deckhand to bosun to first mate of *Andromeda*, all the while cementing a relationship with the owner. Perkins regarded him as quick, responsible, and loyal. Some crew came and went—they were entirely free agents, and between better offers from other boats and better offers on the personal side, turnover was significant—but it was unlikely that Gartner was going anywhere. "I liked to grow my own captains," Perkins said, echoing the view he'd long had in the venture-capital business that he preferred to create his own entrepreneurs rather than accept them at his doorstep after they'd been fully formed. That meant ensuring the captain took the right safety and engineering courses—and Gartner wound up receiving enough training to skipper anything smaller than a cruise ship—but more important, it meant getting a captain to adapt to his idiosyncrasies, not the least of which was that Perkins liked to drive. On most yachts, the owner was content to sit on the aft deck and entertain guests, ruling the bridge from afar. Perkins spent hours at the helm, fine-tuning the sails and running the show up close. For the captain of that vessel, it dictated a lesser role. For Gartner, being first mate of *Andromeda*—the No. 2 officer on board—wasn't bad, but it was likely to be a long time before he ever had the chance at command. Perkins knew he was restless. In time, he sent Gartner on to *Mariette* as the engineer and then as the captain, and then on to the *Maltese Falcon*. Along the way, though, Gartner's woes with women continued, and his legend grew.

Depending on whose stories you believed, he seemed to have girlfriends in many different ports, two girlfriends in the same port, and any number of distant admirers on whatever boat he was working on. Perkins told the story of when they were leaving Auckland on their round-the-world trip, Gartner was saying goodbye to one girlfriend at the bow of *Andromeda* and another at the stern. As the boat left the dock, the two women apparently saw each other, at which inopportune time Gartner offered an enthusiastic "Ciao!" "He picks up women like a blue suit picks up lint," Perkins said with both awe and dismay. Many of the tales were probably exaggerations, which embarrassed Gartner, though not to the point that he discouraged it. " I am a sailor, after all!" he loved to say. But the saga of Debbie (not her real name) was the one that all who knew Gartner—and adored him—relished most.

Long after his status with Perkins was solidified, Gartner got himself a condominium in Antibes, France. It was a place to send mail, keep belongings, and call home. It was conveniently located on the Mediterranean, where Perkins's yachts spent considerable time and, in the middle stages of the *Falcon* project, it was a good base from which Gartner could stay in touch with Gerry Dijkstra, Damon Roberts, and Ken Freivokh. Gartner's American girlfriend, Debbie, was living with him there. To say that she and Gartner had a stormy relationship was like describing Arab-Israeli relations as up and down. During one of their low moments, Debbie locked Gartner out of the apartment and threw his TV, stereo, and clothes out the window onto his car, which turned out to be especially inconvenient because the keys to his car, as well as to the apartment, were still inside the building. There was a scene, the police arrived, and Gartner was arrested for creating a disturbance. (The Perkins sailing team just didn't make out well with the French gendarmes.) Perkins was in town soon thereafter and took Gartner to dinner. "You've got to get this woman out of your life," Perkins advised him.

Gartner agreed to disentangle his affairs from Debbie. Yet six months later, at a big regatta in Italy, he announced that he and Debbie were back together. "We've worked it out—she's the love of my life." Gartner asked if she could be a stewardess, perhaps on *Mariette*. Perkins said no, but offered that she could stay with Gartner on whatever boat he ended up on, as his companion. There were of course husbands and wives aboard superyachts, and there were lesser degrees of romantic entanglements, however much owners and captains preached against them. But inviting on a girlfriend in a more or less permanent context was asking for trouble, even if it wasn't Debbie and Chris.

The arrangement didn't last long enough to make for mere trouble. When Perkins arrived back from a regatta dinner, he said, the boat actually was shaking. The putative couple were throwing things at each other, hollering throughout. By the next morning, Debbie had walked off. Humiliated, Gartner gave Perkins his letter of resignation at breakfast. Perkins tore it up, but insisted that Gartner rid himself of Debbie once and for all. They bought her a one-way ticket back to the United States and Gartner went to stay on a different yacht temporarily. Debbie was back in no time. "Where's Chris?" she asked. "I love him!"

"He's gone, and so are you," Perkins told her. It was four in the afternoon. Her flight was at seven. Perkins got a driver to take her to the airport, telling him, "You must make sure she's on that flight." It was like a poorly scripted episode of *24*. At the airport, Debbie excused herself to use the bathroom. The driver never saw her again. He assumed she got on the flight and that he just missed her. At eleven p.m., Debbie showed up at the boat—"the devil incarnate," according to Perkins. She was all dressed up. Perkins saw her coming and scurried Gartner off again. Debbie did not take the ploy well. When she couldn't find Gartner on *Atlantide* or any other boat—when she heard the *Atlantide* crew partying, she assumed Gartner was at the center of it—she went ballistic. She threw guests' shoes in the water and she wound up in the drink herself. But the next day, at long last, she flew home without further incident.

As entertaining as the Debbie tale was, it didn't much endanger shipboard harmony. The Nympho Chef was another story. A stunning blonde on one of Perkins's boats before the *Falcon,* she was supposedly married to another member of the crew. The problem was that she was also having an affair with the first mate—as well as an engineer. And the unwinding of that sexually combustible disaster came while the boat was in the middle of the Indian Ocean. When they reached Fiji, Perkins put the chef on a plane off the island. As it turned out, they learned that the chef and the crew member were not married. When he mentioned the mess to Gerd later on, she told him she knew all about it from her times on the boat, adding that the chef wasn't ultimately interested in the captain or the mate or any of the crew. According to Gerd, she wanted the owner. In the world of crew politics, that was the thermonuclear threat—though from the perspective of the pursuer, it was the pinnacle in career advancement.

Even so, having the correct chef was vital. Having tasty vittles could mean the difference between an enjoyable voyage or a miserable one. If the food stank, there wasn't a Michelin-rated restaurant around the corner. If there was no food at all, that was a calamity. That happened aboard the occasionally woebegone *Mariette*. In a 1997 trans-Atlantic race to Falmouth, England, the chef failed to adequately provision the boat. The main reason was a bad navigator, who believed that barometric-pressure readings were the way to plot the best direction, rather than using the tried-and-true straight-line method. The navigator, of course, was a Frenchman, Perkins was quick to point out. The novel approach caused the boat to zig and zag way off course. On the fifteenth day of what was supposed to be a twelve-day trip—still two hundred miles from Europe, in cold and rainy conditions—*Mariette* turned into a version of *Lord of the Flies*, with crew fighting over M&Ms and a box of Chips Ahoy. Before the boat left New York, Perkins's fiancée Danielle Steel—no ocean traveler herself—had continually asked if there was enough food aboard. Perkins replied that he didn't have to micromanage such things the way he said Steel micromanaged her own affairs. Steel has laughed about the incident the whole time since.

Once the construction of the *Maltese Falcon* was fully under way, Perkins asked Gartner to move to Turkey to stay on top of the project. Perkins flew in every six weeks or so, but he wasn't about to move in at the yard. Gartner, on the other hand, rented a house just down the road—and the women of Tuzla got a new friend. As the boat moved closer to completion over her last two years, engineers and stewardesses were brought on, and they, too, made Tuzla their home.

By the early spring of 2006, the project was two years late. Preparing for his new nautical creation, Perkins had long since sold the *Andromeda la Dea* ketch and in the year before bid farewell to *Mariette* as well. Only *Atlantide* was left in the fleet and he had brought the motor-sailor to Tuzla to live on during his trips to the Yildiz yard to oversee construction of the DynaRig. During that period, he spent extra time in England at Plumpton Place, which was only a few hours from Turkey by air.

In early April, the big boat finally emerged from her specially built shed, through giant doors. "Aman Allahim!" gasped several of the Turkish workers, which translated into a good old-fashioned "Ohmigod!" Baki Gökbayrak wept, as if seeing one of his children go off to college abroad.

On wooden greased skids, over the course of two days—the slowest she'd ever move, Perkins hoped—the *Falcon* inched her way onto a dry dock that a week later was towed out into the deeper water of Tuzla Bay and then submerged, leaving the clipper yacht afloat for the first time. When it did float—not that anyone expected the laws of physics to fail—everybody watching from on board *Atlantide* cheered: Perkins, Gerry Dijkstra, Fabio Perini, Damon Roberts, Ken Freivokh. "It floats, it floats!" cried Freivokh in good fun. Dijkstra, in particular, was relieved that the *Falcon* sat in the water where it was supposed to—not too low, as he feared, despite the fact that every component of the boat was weighed before it got added to the hull.

The following Sunday, Perkins threw a huge christening party, which all the workers and their families attended. For most of them, it was the most lavish event they'd ever witnessed, but what especially got their attention was the boat itself. The wives and young children kept going out to the dock to see what their husbands and fathers had helped to make. Perkins couldn't resist a little mischief. In a speech to the workers, Perkins took a minute to praise "my friend Dr. Jim Clark's *slightly smaller yacht*" as an example of European craftsmanship. Clark's crew got wind of the comments by the end of the day. But they didn't tell Clark.

Without her sailing rig, the *Falcon* looked strange, like a motorboat gone terribly wrong. Soon enough, though—in another week—two five-hundred-ton cranes from Russia arrived by barge down the Black Sea to step the towering, massive masts into the boat. With thirty men on the dock and on deck guiding the cranes, the masts finally became vertical and gave the *Falcon* a scale to equal the freighters elsewhere in the port. Early the next morning, Perkins—ever the lathe operator and would-be TV repairman—was atop the mainmast, inspecting the electrical connections and showing the Turks he was as hands-on an owner as they had ever seen. The *Falcon* still looked very odd and nothing like the *Cutty Sark* or a schooner or a ketch or anything else on the water. A major yachting magazine, on its cover, called the *Falcon* "the biggest, strangest sailboat ever." Its masts were thicker than on any other boat, its yards extended way beyond the beam, the entire rig seemed out of proportion to the hull attached to it. But at least you knew it was a sailboat. In a tradition dating to Roman times, Perkins put a coin under the mainmast—a gold piece with the likeness of Atatürk, the founder of modern Turkey. The crew insisted on it, for the superstition was that if a vessel met her doom, at least this way the dead had compensated the gods for passage to the afterlife. For a seasoned venture capitalist, it seemed like a reasonable price to pay.

TWELVE

Mine's Bigger, Mine's Better

The route out of Istanbul by water is southwest through the Dardanelles, the thirty-seven-mile-long strait that connects the Bosphorus
and Sea of Marmara with the Aegean and Mediterranean beyond. The
ancients called the narrow waterway the Hellespont: Troy once existed
near its Aegean entrance and the Trojan War took place on its Asiatic
shores. In more modern times, the Dardanelles were a strategic target of
various warring countries—from its defense of Constantinople, to its
blockade by Russia in the Napoleonic Wars, to its loss to European interests in the Crimean War in the 1850s, and to the killing fields of Gallipoli
during World War I.

The *Maltese Falcon*'s maiden voyage certainly wasn't in those leagues.
Yet, after the big party at the Çiragan Imperial Palace and the fireworks
display the night before—as the clipper yacht unfurled her sails and
approached the Dardanelles—Perkins couldn't help but take note of the
location and why this was a momentous day for the technology of sailing.
Here, he said, was a modern square-rigger, "the likes of which the world
had never seen, passing through waters that have seen a lot of history." At
the palace, Baki Gökbayrak, a student of Turkish history, had wondered
aloud if the yacht he helped to build, as it passed through the Dardanelles,
would begin to "fulfill her destiny as the greatest sailing vessel ever built."

As the *Falcon* left the country of her birth, Perkins was downright wistful about the five years it took to spawn that creation. "If you thought I was always certain this would happen, you misunderstood my confidence," he said.

Lowest on the masts, the first three sails to set—the two courses and the cro'jack—rolled out of their respective mandrels effortlessly. It was an operation that lasted just over a minute and was eerily silent—the hum of the unfurling motors easily drowned out by the wind. But the next sail, the lower topsail on the foremast, jammed. In a building breeze of twenty knots and on a broad reach—with the wind coming at the sails from an angle just aft of the beam—that was not an auspicious sign. If the sails were this finicky, all that testing and engineering on the DynaRig would have wound up having been a waste of time. In the wheelhouse, Perkins grumbled, but his Dunhill hat remained on his head and there was no hat-stomping in these first hours of the *Falcon*'s maiden voyage. At the very least, it was a good thing that the sail got stuck when we were well out of sight of most of the tourist fleet that had been watching in and around Istanbul.

Christian Truter, the spar master, went up the mast, fixed the tensioning problem, and the lower topsail emerged. Up high on the rig, he looked like a mere fly on the sail. As it turned out, though, it was one of only three times that Truter went aloft during the ten-day trip in the early summer of 2006. Given that the sails were successfully furled and unfurled in 380 round-trips over the course of the shakedown cruise—in all different kinds of wind conditions—Truter had virtually nothing to do. He was in fact disappointed, as he enjoyed the ride most from 190 feet above the waves. Frankly, the entire crew was grateful to finally be out of the Tuzla shipyard. It was dirty, noisy, and worst of all, it was on the shore— the crew, like Perkins, longed to be at sea, "to let the *Falcon* spread her wings," as Sam Bell, the third engineer, put it.

"The engines went off and away she goes, throwing water off her shoulders," wrote George Gill, the poetic second engineer from Britain, who kept an online log. "If smiles could be measured in lumens, there were thirty-three people aboard who could have lit up Twickenham!" referring to the famous rugby stadium outside London. With all fifteen sails successfully deployed, we were romping along at fourteen knots. A complete tack took fewer than two minutes, with much of that interval devoted to refilling the sails with wind. That was almost ten minutes faster than Joe Vittoria's *Mirabella V*, and faster, too, than Jim Clark's *Athena*. Jibing on the *Falcon* was almost imperceptible, as there were no booms to whip across the deck; the masts merely had to rotate twenty or so degrees, with the attached yards and sails moving around with them.

Sitting in the aft cockpit area—right behind the Sinnex doors—we knew the speed every second. Perkins had installed a red digital readout at the base of the mizzenmast that displayed the boat's speed. It was like watching an electronic stock ticker on Wall Street. Faster, faster! In addition to Perkins, those aboard were the sixteen crew members, six technicians from Perini Navi, two naval architects from Dijkstra's firm, Roberts, Perkins's son Tor, three couples who were among Perkins's closest friends, and me. Perkins toasted Perini, Dijkstra, Roberts, and himself. "Sometimes," he said, "the geeks win!" By dinnertime, Perkins consented to reduce sail area to give the crew some rest and to make the boat more comfortable for sleeping after a hard day. In ten knots of wind, the *Falcon* was still doing seven knots. But comfort and ease were not what Perkins had in mind for most of the maiden voyage.

At 3:45 in the morning, I got rolled from my berth. Nobody had told me that my big queen bed had leeboards—a wonderfully ancient device designed to prevent sudden humiliation and an instant concussion. A leeboard was a piece of wood on the side of a bed—typically stored underneath the mattress when not in use—that prevented the occupant from spilling out when the boat heeled. When I went to bed, the *Falcon* was heeling at a modest ten or eleven degrees as we emerged from the Dardanelles into the Aegean Sea. This was just fine, and was typical as we sailed on. It wasn't like sleeping at the Ritz—call me inflexible, but I've always had a thing for sleeping on a level surface—but this was a boat and I made allowances. Even the sound of the swells outside my window was calming and more than compensated for the hydraulic creak of the main-mast, just up the hallway from my stateroom, in its frequent rotations during sail trims.

But much past eleven degrees, a leeboard was a good plan. Past fifteen degrees, it was a great plan. The hull's relative slenderness, when com-pared to her length at the waterline, meant the *Falcon* was quick to heel. The boat also heeled more because the hull was flared—from the deck line down to the waterline, the hull tapered at a relatively sharp angle. The combination of narrowness and flare made the *Falcon* more efficient through the water, but it was less buoyant as it leaned over. The typical Perini was "fat and flat," which made it more stable, but also slower.

At fifteen degrees, I became a projectile. At the velocity at which I barreled into the wall, I figured that Columbus was wrong and that we'd sailed off the edge of the planet. Curious to know whether it was time to e-mail goodbye to my family—alas, the Internet wasn't working, but that was a different issue—I put on a pair of pants and maneuvered up the two flights of the atrium to the wheelhouse. At a severe angle of heel, the atrium was an adventure. Its circular staircase meant that for a few steps you did feel like you were ascending. But as you went around, you leveled off and were actually then going down—with Lorraine Shemesh's jumbo swimming-pool painting staring you in the face—and in a few steps you were then headed back up. There was a handrail along the wall, but it was pretty much beside the point. Falling over was your second worry. You felt like you were still in one of those amusement-park funhouses of strange doorways and mirrors, where up was down and your middle ear at best provided only hints. I wasn't going to heave in Perkins's incredible yacht, but then again, nobody told me to bring along an emergency purse just in case.

Whom did I find in the wheelhouse, other than the two engineers on watch? It was Perkins—in his jammies—at the sailing-control station. As was often the case, he couldn't sleep, so he got up from the adjacent passage cabin and came into the wheelhouse to play. Under the light of the moon, with only a few tankers on radar, he had unfurled four more sails in the freshening early-summer breeze—now at twenty-two knots—and more than doubled the *Falcon*'s speed to fifteen knots. "Isn't this great?" he marveled. The only problem was that as we were sailing a little closer to the wind and with a little more sail, the boat was going to heel more. We had gone past fifteen degrees—what Perkins had said was comfortable and safe for people walking through the boat—and all the way to twenty-four degrees, which was more than a quarter of the way to the boat being flat on her side. Well, nobody was going to be walking around at that hour, right? *Wheee!* It was on the way to twenty-four degrees that I exited my bed (and saw a big leather chair do a back flip). I imagined I now knew what it was like looking down from the top of Everest. As I learned thereafter, a few other guests also went tumbling. Much of the crew slept with leeboards regardless, and half the crew didn't have a problem this night because their berths were built right against the side of the boat that we were leaning toward anyway. Never mind that Perkins had said many times when the *Falcon* was being designed that "I won't let my boat heel past fifteen degrees." The reason: "It's too scary for everybody." Except him.

After some more fun at this heel and speed, Perkins went back to bed. Someone else on watch furled several sails and we returned to a more civilized angle of heel. Until breakfast.

———————

A few hours later, at 7:45 a.m.—right around the time his soft-boiled eggs and coffee were being prepared—Perkins wanted to test out the crew in real-life conditions. The DynaRig—the high-risk proposition of the *Maltese Falcon* project all along—was behaving fine. Now it was time to discover how mealtime was going to work out leaning hard over to starboard. Perkins set a few more sails. And then, without notifying any crew, apart from those in the wheelhouse, Perkins steered the boat hard up to the wind, so the angle of heel suddenly went from twelve degrees to twenty-two degrees. The cap rail on the leeward side was buried underwater—making that side of the deck impassable, unless you were a fish. You could hear people's cries of surprise from around the boat and they weren't saying, *"Wheee!"* On the crew walkie-talkies came the sounds of crashing plates and clanging glass in the galley, three levels below. "What the hell is he thinking?" hollered Harriet "Harry" Davies, one of the stewardesses. "Why didn't Tom warn us?"

That was too much for Perkins. "This is a sailboat!" he proclaimed to anyone who would listen in the wheelhouse. "A sailboat heels! It sometimes heels without warning!" Off came the hat and onto the floor it went, stomped upon by both of Perkins's sneakered feet. This was the famous—or infamous—hat-dance. This was a moment when a mere sigh just wouldn't do. "Fuck!" he said three times, with an emphasis on each exhortation. "We just need to be better prepared when we're sailing!" Several of Perkins's friends made it out of their staterooms and up through the atrium. When they learned that the boat wasn't in trouble—and that Perkins was at the controls—they just smiled knowingly. They had sailed many times with Perkins on his other boats.

On the aft deck, under beautiful blue skies, Truter was trying to secure the furniture—rearranging the deck chairs, as it were. With the waves washing in, some of the furniture was about to go in the drink. "I'm so glad I scrubbed the salt off the deck earlier this morning," he said. Jed White, the chief engineer, got "head-butted" by the German engineering manual in the engine room. "When Tom's driving the boat," he said, "we have to hold on for dear life." Added Rob Bell, the first mate, as he directed deckhands to go through the boat and make sure anything loose was tied down: "It's his boat, after all." None of the crew was complaining. They were only describing the facts of life aboard the *Falcon*, but they were also doing so with a degree of admiration that the owner for whom they worked was, on occasion, a madman.

In the galley, the assistant chef, Natalia Singleton, mourned the loss of yet another set of eggs for the owner. As high-tech and decked out as the *Falcon*'s infrastructure was, the stovetop and ovens contained no gimbals—simple mechanical devices that allowed an object mounted within it to remain in a horizontal plane despite external motion. Ships used gimbals for compasses and weather instruments on the bridge, and chronometers and lamps in living quarters. And in the galley, gimbals allowed the safe boiling of water, as well as the even cooking of food. But on the *Falcon*—the postmodern clipper yacht with a radical, utilitarian rig—a catering consultant had decided that gimbals weren't aesthetically pleasing. (*Mirabella V* didn't have gimbals either, but Vittoria said the chef was always given a heads-up about the boat's sailing plans for a given day.) Singleton therefore became adept at holding a pot of water just so while under sail. And with Perkins's eggs, if he happened to want them sunny-side up, she had learned to position the pan to make sure they came out right—until the boat heeled more, and the eggs slipped the surly bonds of the pan. One time, she called Rob Bell in the wheelhouse to ask if the boat couldn't lose a sail or two just so she could make breakfast a bit sooner. Bell told her that was not the Perkins way. She found the whole exercise hysterical.

Almost as amusing to Singleton—and everybody else—were her "starboard-tack muffins." While redoing a plate of eggs took just a few minutes, blueberry buns required work. There was no way to adjust the batter in a muffin tin, so she under-filled each compartment, so that when the boat heeled, the batter didn't run out. The result, though, were muffins that listed. Her starboard-tack muffins (or "port-tack muffins," if you were so inclined) could just as well have been called fifteen-degree baked goods. Either way, she was tempted—literally—to invite Perkins down to the galley some time to see what his ship design had ironically wrought: twenty-first century rigs on deck and eighteenth century food preparation down below. There was also the long, adjustable nozzle for the main sink. It looked great, but when the boat heeled to port, the nozzle leaned against the splashboard, and the running water filled the countertops and flooded the floor. "So many of the things on this boat were designed for aesthetics rather than practicality," she said. But she decided against inviting Perkins down. When he did visit now and then, she was happy just to give him a bowl of Ben & Jerry's Cherry Garcia and leave it at that.

————————

Negotiating the staircase, remaining in bed, cooking in the gimbal-less galley—these were challenges. But nothing matched the terror of the bathrooms in the *Maltese Falcon* staterooms. The world's finest-looking marble—in this case, black Portoro edged with cream-colored Novona Travertine—was all well and good, but when it got wet, it was slicker than an ice rink. Keep the floor dry and you were fine. Get it wet, and unless you emerged from the shower in scuba flippers, you were going to have some difficulty. Add in some boat heel in the wrong direction and, well, you get the idea. For the first few days of the maiden voyage, the floors were distinctly un-dry. The problem was that the artfully designed drain in the shower—not an ugly metal strainer in the center, but a recessed rectangular channel around the edges—didn't seem to work when the boat was tilted. If the boat was heeling in the direction of the shower bench, that was fine—the water just swished around. But if the boat was heeling in the other direction—as it was in my room for most of the initial one thousand-mile leg across the Mediterranean—the water escaped underneath the glass doors, coating the Portoro and Novona Travertine floor, before piling up on the other side of the bathroom. The water would've made it under the bathroom door into the carpeted bedroom, but fortunately all the plush cotton towels I'd hoarded formed a nice dam.

I mentioned all this to the engineers, who knew about the issue from the other staterooms. The fix would have to wait. My showers would not. That was not necessarily the right call, leaving aside that the hot- and cold-water dials seemed to be only a suggestion. On my second day, as the boat was heeled well over, my feet gave out on the slippery floor as I got out of the shower. I managed to break my fall with my hand, but my face still collided with the bidet. My first reaction was that I might've split my head open. My second reaction was wondering if blood could easily be removed from porcelain and marble before I passed out. I had grown up sailing on Long Island Sound, but had never been well out to sea. So I had my small fears about freak waves, white squalls, pirates-pursuing-Perkins, and what happened if you got appendicitis in the middle of nowhere. But never did I think of meeting my demise in the head. When I informed Perkins and Gartner later that morning of my fall, I suggested that, had I perished, they'd grieve for ten minutes and then bend over in laughter at how it happened—the stupidest slip-and-fall in history. They told me I was exaggerating: it wouldn't have taken a full ten minutes. Perkins did ask if there was blood on the floor. I'm fairly certain he was kidding. It helped matters that I had been considerate enough not to bleed.

Then there were the custom planar sinks, carved out of blocks of limestone. They were gorgeous—and sinister. The sink was indistinct from the long countertop—a neatly angled rectangular indentation below the faucet. You turned the water on, you washed your hands or brushed your teeth, and the leftover water hit the limestone slab and flowed to the edge of the indentation and then disappeared over the edge, into a trough, to a concealed drain well below. It was a work of art, worthy of display in a sculpture museum. But it was unusable. The design defect was that the surface was neither slippery nor slanted enough, so your Crest residue never made it down the slab. The water did—except when the boat was heeling, which it usually was, in which case water flowed not to the drain, but over the edge of the indentation, onto the countertop, running over it, and streaming into the towel dam on the floor containing the water from the shower. "Even for me, the water won't go uphill," Perkins acknowledged. The sink drain—such as it was—leaked, too.

Jerry Di Vecchio, one of Perkins's friends aboard, gave up on the sinks. She announced at breakfast one morning that she had taken to brushing her teeth in the bidet, which was brand-new anyway. It seemed reasonable under the circumstances. And after all, she was the food and entertaining editor at *Sunset* magazine, an arbiter of good taste. Perkins chortled about the bidets: he hadn't wanted them to begin with, but was convinced they were necessary if he was going to charter the boat. He laughed less about the rest of everybody's bathroom travails, which were solved in due course by repairs to the drains. For the sinks, Perkins and Freivokh, the interior designer, eventually installed aluminum inserts that solved the runoff deficiency. Neither Di Vecchio nor anyone else ever had to brush in the bidet again. In his own defense, Freivokh—who prided himself on the meticulousness of his designs—said the showers and sinks had not been built precisely to his specifications. It was difficult to tell, but the bathrooms did recall Fabio Perini's observation that "everyone's *design* project is always perfect." What looked good on paper didn't always turn out so great in three dimensions, though the bathrooms were the only part of Freivokh's plans that needed retrofitting. The other unusual features—the cantilevered steps of the dramatic atrium, the dining-room skylight, the spider struts supporting the mizzenmast, as well as the Sinnex doors of the saloon—worked flawlessly.

And when something did go wrong—not because of design but malfunction—the crew tended to have a sense of humor about it. During the first charter of the boat, later in the summer, one of the pipes in the ceiling of the main deck, which supplied water to the stateroom bathrooms below, burst. That produced torrents of water cascading through the atrium and accumulating on the lowest level. Before it was fixed, the crew went through dozens of towels and filled many buckets to contain the mess. At €335,000 a week, it was rather embarrassing, but the charter guests took it in stride. It helped that the crew called the leak "our special water feature." It was probably a good thing that Perkins wasn't aboard at the time.

———————

Through it all, the DynaRig worked to perfection—almost anticlimactically, given the anticipation over its first extended use. "We'll know in about five minutes whether we have a good boat," Perkins remarked with gallows humor when the sails emerged as we left Istanbul. "If not, then, well, we'll have some great motor yacht." He knew from a day of sea trials during the prior month that the electronics all worked. But that was just a basic test to see if wires and the embedded fiber optics were connected, if motors responded to on and off switches, if the masts rotated when asked, and if the sails went in and out. By contrast, the maiden voyage was designed to push the *Falcon* hard, and to subject her rig to repeated and varied stresses. And the DynaRig performed brilliantly. Perkins was exhilarated—this was what he had worked so hard for.

Each time Perkins set more sails or each time he brought his yacht closer to the wind—and adjusted the sail trim by rotating the masts accordingly—the *Falcon* responded. When he sent the boat over to twenty-four degrees in the middle of the night, the bending-and-torque screen at the sailing-control station indicated significant stress on the masts—two-thirds of the maximum that Roberts and Dijkstra had initially calculated to be safe—but those forces were expected. And while the boat briefly slowed down during a tack, it was behaving just like any other sailboat that was beating to windward. With even more sails up, the DynaRig would have been just fine, even if Harry the stewardess and the rest of us might not have been. Gartner still didn't like the idea of sailing a boat from indoors, but he understood that the absence of a flying bridge was one of the tradeoffs in designing the DynaRig and he was getting used to it. In short, the *Maltese Falcon* was showing itself in these first days to be a smashing success—and Perkins basked in it. He didn't sail the boat continuously—and when the wind occasionally died and the engines had to be turned on, he lost interest entirely—but he could be found in the wheelhouse periodically throughout the day and much of the night.

His biggest worry had nothing to do with his new clipper yacht, but his land-bound boardroom struggle back in northern California, at Hewlett-Packard. A month earlier, he had quit as a director of the company he had long loved and had been associated with for half a century. In a tempestuous board meeting in Palo Alto, after learning that HP chairman Patricia Dunn had approved surveillance on board members without their knowledge, Perkins resigned on the spot and marched out of the boardroom. The spying included obtaining the personal phone records of the company's own directors to root out a leaker at the highest levels of HP. The company did so by using private investigators who engaged in "pretexting"—calling up phone companies and impersonating directors seeking their own records. Perkins was enraged at the invasion of privacy, all the more so when the surveillance pointed to a close friend of his as the leaker.

HP announced Perkins's resignation in a perfunctory press release, but gave no hint of the boardroom turmoil. Meanwhile, he stewed. He risked legal exposure if he went public about confidential corporate proceedings, yet he had potential liability as a director during the time of the spying. He also came to believe that federal law required the disclosure to securities regulators of the circumstances of his resignation—a public filing in Washington that in turn would generate rabid press coverage, given the spying context and the *dramatis personae*. When he wasn't in the wheelhouse of the *Maltese Falcon*, Perkins often retreated to the passage cabin to fire off rounds of e-mails to Dunn, other members of the HP board, the CEO Mark Hurd, and everybody's lawyers. Thus, in one room of his boat, Perkins sat at the helm of a magnificent sailing machine—the product of a lifetime of sailing passion and venture-capital booty. But in the room next door, he was being sucked back into the world that the boat was supposed to provide the ultimate escape from.

Worst of all for Perkins, who understood and appreciated irony, it seemed that he might actually be enjoying the action in both rooms. Being in the maelstrom was fun either way—at least a little bit. In mid-July, he was putting on a three-day lavish "international debut" party for the *Falcon* on the Italian coast, near Perini Navi's headquarters. Rupert Murdoch, Jim Clark, Danielle Steel, the Italian prime minister—the guest list was impressive. But hours after Perkins had quit the HP board in May, he e-mailed Dunn:

> Hi. In view of today's events at HP, I would appreciate your considering my boat party invitation as "not sent." Thanks Tom.

It was a bucket of ice-cold sea water in the face of Dunn, who had long been insecure around Perkins. While she was a successful executive at an asset-management division of Barclays in San Francisco, she was a small fry in the zeitgeist of Silicon Valley. Despite being appalled at the e-mail, Dunn graciously agreed not to attend. Three days later, Perkins e-mailed again, with "one last bit of advice."

> Given last Thursday's debacle, you should resign as chairman of the board . . . Now that the dust has settled, I can only paraphrase Nathan Hale: I regret that I have but one HP board seat from which to resign.

Dunn let that suggestion pass, at least until the scandal became public and the board of directors ousted her.

Perkins was pleased with both of his e-mails. It was part of the take-no-prisoners attitude he had used in venture capital and in his personal affairs, whether with the chairman of Morgan Stanley over $100,000, or the hapless Turkish customs officials a few days before. And Perkins lost no sleep when Dunn was forced out of HP later in the summer, though he was sorry to see her indicted by the state of California—and to see himself painted in some press reports as a bully, especially in light of her bouts with cancer. (The criminal charges against Dunn were dropped in the spring of 2007.)

Yet even with his evident triumph, Perkins came to wonder if he had overreacted in the HP matter. He dearly missed serving on the board, whether it was discussing competitive strategy with the CEO or operations issues with engineers in the trenches. Had he sacrificed his position in favor of principle or had he indulged a fit of pique? It was a kind of self-doubt he didn't share with many people. But at the keyboard of his laptop in the passage cabin, as he agitated to make the scandal right, he was anguished by the notion that, as a personal matter, he might have gotten it wrong. In the end, did he conclude he'd made the right call in going to war with Dunn? It was impossible to tell. Depending on his mood, he might answer either way.

———

Sailing westward across the sea, we reached Malta four days later. Perkins intended the country—centrally located in the Mediterranean—to be the home port for his yacht someday. Lining the shores and waiting at the dock of this tiny but physically dramatic nation, the locals treated his arrival with fanfare that made him momentarily forget the brewing Hewlett-Packard storm. A resident sailing historian lauded the *Falcon* as "the most important sailing innovation since the clipper ships themselves." Malta's fortressed, deep-water harbor was one of the best in the Med, but it was also narrow and any large vessel had to take aboard a government pilot to steer it through the channel. On our departure, the pilot on the *Falcon* was more agog tourist than navigator. He not only let Perkins stay at the helm as we motored out of the harbor, but he allowed him to put up the sails to give the waving crowds the chance to see the boat in full regalia. This was the kind of treatment that Perkins adored.

It was much the same scene as we went through other historic waterways—the Strait of Messina, between the eastern tip of Sicily and Calabria; and the Strait of Bonifacio, between Corsica and Sardinia—on our way to Antibes and the French Riviera. Party boats, day sailors, other superyachts—and one warship from the U.S. naval base at La Maddalena, Italy—wanted a look as Perkins's yacht cranked along. Along the way, we sailed by the *Wind Spirit*, a 440-foot, four-masted, motor-sailor cruise ship, whose captain radioed to the *Falcon* wheelhouse, "What kind of vessel are you?" and "Why do you have your engines on when you are sailing?" The *Falcon* crew laughed, for we did not have our engines on—and hadn't had them on for most of the journey. As the *Wind Spirit* came into view, Perkins ordered all sails up, just to make sure we raced by at full bore.

Later, as we drew closer to France and the end of the 1,600-nautical-mile maiden voyage, an e-mail arrived from the captain of *Ronin*, a huge powerboat owned by Larry Ellison, the Silicon Valley tycoon behind Oracle Corporation. The captain had a charter guest who was eager to see the *Falcon* and get together with Perkins. *Ronin* was 350 miles away. The guest was Steve Jobs, along with his family. Perkins replied back that he'd be happy to see Jobs—if *Ronin* could catch up. In Antibes, we were able to pull into the largest slip in the largest yacht marina in the Côte d'Azur—the place that Paul Allen normally kept one of his megayachts. Among hundreds of other boats, the *Falcon* towered over the rest—so much so that the crew on the even larger powerboats in the harbor came by continually to ask for a tour. When I flew back to the United States a few days later from nearby Nice, the Delta pilot dipped his wing as we passed over the harbor; his photo of the *Falcon*, from twelve thousand feet, appears on the *Falcon's* Web site.

———

At summer's end, Perkins finally got a chance to measure his boat against one of the other two megayacht sailboats. At the Monaco Boat Show in late September, the *Falcon* was the belle of the ball, with Prince Albert in attendance, and enough barons, knights, lords, countesses, duchesses, and earls for a royal wedding. Right next to the *Falcon* was Joe Vittoria's *Mirabella V*, its 292-foot single mast dwarfing the *Falcon's* DynaRig. On the morning after the last day of the show, the two owners agreed to engage in a "friendly competition that involved going as fast as possible next to each other"—with a Discovery Channel helicopter filming and scores of boats watching. Perkins happily called it a race. Vittoria could not—for his insurers forbade *Mirabella V* from competition. Both gleefully noted that Jim Clark and *Athena* were halfway around the globe, on the way to New Zealand.

A dozen miles beyond the jetty of the Monaco harbor, the *Maltese Falcon* and *Mirabella V* went at each other. With a twenty-knot breeze kicking up to twenty-five out of the northeast, *Mirabella V* set her full mainsail, along with a big jib and a staysail. With her designer Ron Holland aboard not uncoincidentally, the sloop was as powered up as she'd ever been. On the *Falcon*, at the helm himself, of course, Perkins had all fifteen sails up—the cap rail in the water, crew at the ready, and surely no eggs or muffins cooking in the galley. Sailing close to the wind, each boat held its own, each hitting fourteen to fifteen knots—*Mirabella V* going a little faster. Even the *Falcon*'s side decks were well awash as the boat neared a twenty-degree heel, but Perkins was getting no additional drive from it, so he furled the mizzen royal (the sail furthest aloft on the aft mast) and lost no speed at all. His boat wasn't outgunning Vittoria's, but beating to windward was supposed to be the sloop's strongest suit, and Perkins was thrilled. Despite the overcast sky, the *Falcon* and *Mirabella V* together were breathtaking as they ripped through the rising seas, sometimes passing within fifty yards of each other—a shadow dance among two of the three largest megayachts in the world. A *Falcon* engineer with a sense of humor thought of snapping a photograph to e-mail off to a yachtie friend on *Athena*.

After nearly an hour off Monaco, the two boats bore away from the wind for a speed run. The *Falcon* surged to seventeen knots and pulled ahead. *Mirabella V* did almost as well—until a gust of thirty-five knots overpowered her mainsail and put a fifteen-inch tear in it, forcing the boat back to shore for quick repairs. Though he'd long been skeptical about *Mirabella V*'s practicality—and this minor mishap seemed to prove the point—Perkins didn't gloat at all. As we set out for Cannes on the coast, he was too busy thinking up plans to make his yacht go faster: altering the respective angles of the masts to better harness the wind; modifying software to allow the hydraulic motors to rotate the masts four times faster during tacks; and rotating the masts during a tack only after the rudder had been turned and the bow had moved through the wind onto the opposite point of sail. The last trick was counterintuitive because the back-winding on the sail briefly slowed the boat down. But it also got the bow around more quickly, which more than made up for the loss in speed.

———

Bigness in megayachts can be fleeting. In late 2006, the German shipyard Lürssen had launched what it described as a 305-foot schooner—sixteen feet longer than the *Maltese Falcon*. The name of the schooner was *Eos*. While the yard said the owner's identity was confidential, the yachting community knew it was Barry Diller, the innovative American media executive. Diller wanted anonymity—to the extent that's possible in sailing around in a megayacht—but the claim of the boat's builder of "biggest in the world" was exaggerated. Like *Athena*, whose hull it resembled, *Eos* had an enormous bowsprit. Without it, *Eos* was still seventeen feet shorter on deck than the *Falcon*, and twenty-four feet shorter at the waterline. Perkins could remain secure that his was still biggest.

On the subsequent trip west—along the French and Spanish coasts, and then onward in search of trade winds across the Atlantic to the Caribbean—Perkins hoped to really test out his "Big Bird." While wind was fickle, a good place to go looking for lots of it was the Gulf of Lion, which extended out from the south coast of France to the border with Catalonia. The Gulf of Lion was infamous for mistrals, which could blow as often as a hundred days a year. The mistral was likely Europe's best-known weather phenomenon. A brutal northerly wind, it typically developed as a cold front above the Alps, which barreled down the Rhône Valley, got funneled into ports like Marseille and Saint-Tropez, and then roared out to sea. Wind speeds in the Gulf of Lion could reach hurricane force. About the only kind thing about the mistral was the blue sky that accompanied it.

Despite its nickname as the "idiot wind"—for what it did to people's sanity—Perkins wanted the mistral and he got it. On a moonlit Monday in the middle of the gulf—on the midnight-to-four-a.m. watch—the wind filled to forty knots, then fifty, then gusted to sixty-seven knots. The seas turned white, rising to thirty feet. It was blowing so hard that a large tarpaulin tied down to one of the tenders on deck flew off. Then a huge wave accidentally triggered the emergency-radio-beacon that momentarily alerted authorities a ship was in distress. You couldn't go outside the boat without getting pounded with spray. On the main deck, under several feet of water back by the mizzenmast, you couldn't go outside at all because of the risk of getting swept overboard by a renegade wave. As fast as all that foaming water escaped through the scuppers in the side of the boat, more seemed to come tumbling in.

With only three of her fifteen lower sails unfurled—the two courses and the cro'jack—the *Falcon* hit a speed of 24.8 knots—more than twenty-eight miles an hour, cresting along the waves on a point of sail almost directly at right angles to the wind. It was faster than the boat had ever gone—and more than three knots faster than its theoretical hull speed. Perkins was in all his glory. Well, almost. He regretted that he hadn't thought of installing the check stays on the masts, which would have made them more stable and allowed him to double the sail area. "I think we had a shot at thirty knots!" Perkins said. In the quest for speed, for the goal of perfection, there was always tomorrow.

The Flying Dutchman

In the early 1980s, Tom and Gerd Perkins got themselves a fluffy brown-and-gray Maine coon cat. They named her Kitty Cat. Even as a kitten, she was mean and aggressive, except when she was merely indifferent. Perkins liked animals, though he preferred dogs, but came to learn that Kitty Cat was special. When Gerd was fighting cancer a decade later, the cat seemed to sense her pain. After one particularly difficult chemotherapy session, Gerd arrived home exhausted, and just crumpled into a chair in the bedroom. She began to cry. Kitty Cat observed this and jumped into Gerd's lap and purred. For the rest of Gerd's life, until she died in the summer of 1994, the cat behaved that way around Gerd. Perkins recalled with wonder how the cat changed, and he had trouble talking about it without crying himself.

After Gerd's death, the cat stayed at the Marin County home, looked after by house staff while Perkins escaped to Plumpton Place. He seemed to be less unhappy in Britain, as Gerd had spent her times in remission there, whereas she was usually in California during relapses. But Perkins missed the cat and the memories it carried of Gerd. In time, he arranged to have Kitty Cat brought to Europe to live aboard the second *Andromeda la Dea*. He liked the idea of a "boat cat," which dated to olden days, when crews could use the help with rodent control. But the reality didn't quite match the idea. Even before making it to the boat on the Italian coast, Kitty Cat got stuck in a scorchingly hot Milan customs warehouse because some bureaucrat didn't like the paperwork. Only the intervention by the chief customs official in Rome got the cat out before she suffocated. After a ninety-minute drive, the first thing the cat did upon setting foot upon *Andromeda*—just as the boat lurched ever so slightly at the dock—was to leap three feet straight up. On her first sail the cat got miserably seasick. Perkins could tell because the cat threw up everywhere, including on herself. Perkins bathed the cat himself; it was unclear who hated it more. On her second sail, she did the same thing. And the third. The cat apparently had a sense of irony: Perkins had flown her seven thousand miles across the American continent and Atlantic Ocean, to keep him company and remind him of the woman he loved, and she had no sea legs at all. Perkins didn't like it when people got seasick—now his cat was?

Kitty Cat was transferred to *Mariette* in France, his vintage schooner, where she did just as poorly. "I just want you to understand that should anything happen to the cat," he told the captain, "heads will roll." The captain wasn't sure if Perkins was serious or not—Perkins probably wasn't either. But the following week, when the cat escaped through an open porthole and went tearing down the dock, the captain and a dozen crew members were right behind. Kitty Cat eventually went off to live in the apartment of a *Mariette* crew member. A year or so later, after Perkins married Danielle Steel, the cat accompanied him into her Pacific Heights mansion in San Francisco. Steel was allergic to cats, so the cat stayed in the security office with Steel's big bodyguard. When Perkins and Steel got divorced eighteen months later, the cat wended her way back to Marin County, where she lived out the rest of her years, almost reaching the age of twenty-three. Upon her arrival at the home where she had comforted Gerd so, after five nomadic years, Perkins was convinced that the look on Kitty Cat's face said, "I'm back! Thank you, God." The cat was a fixture in Perkins's bedroom. "I wouldn't let her sleep on my bed," he remembered with fondness, "but of course in the middle of the night—every night—

when I woke up, there would be Kitty Cat on the pillow next to me, purring away."

After Gerd's death, and despite his brief marriage and continuing attachment to Steel, it may have been the relationship that he held most dear. That wasn't a reflection of his feelings for Steel—which were strong, notwithstanding their mutual inability to adjust to each other's lifestyle— but to the power and longevity of his bond with Gerd. For Perkins, Kitty Cat was a connection to that which he lost when Gerd passed away. The efforts to which he went to try to keep that unseaworthy Maine coon with him, and the sweetness with which he spoke of her, suggested nothing less.

Over the course of thirty-three years, Tom and Gerd Perkins had a vital marriage. She spoke her mind and had independent interests. But he was the handful, by his own account. One time, as board members of the San Francisco Ballet, they took opposing positions on who should be named ballet director. When he got home later that day—he had gone back to the office after the meeting—he found Gerd packing her suitcases. She was strong-willed and had done battle with her husband many times, but this time it was in public. "I voted against you," she told him. "I thought you'd be mad." He was astonished, and reassured her he wasn't mad at all—noting that he had won the vote. While the latter was said in jest, the fact she could think of walking the plank after two decades of marriage because she had differed with him in the boardroom was a remarkable window on just how dominant a persona he could be.

Yet as obstinate as they both were, the marriage worked. Even with Perkins as Silicon Valley grandee and even with his superyachts, Gerd was the center of his universe and her death altered the geometry of his life. Perkins built a memorial to her in her hometown in Norway. As a child, Gerd had played among the ruins of a historic church there, but they had become so deteriorated that they were closed to the public. The community planned to erect a glass-and-steel enclosure to preserve the ruins; Perkins and Gerd had talked of funding the effort. Two weeks after she died, Perkins flew to Norway to deliver a check for several million dollars. Later on, he was knighted by the King of Norway for his benevolence. Today, the "glass cathedral" is the pride of Hamar, Norway, and a plaque inside honors Gerd's memory. At MIT, Perkins's alma mater, an endowed engineering chair is in both their names. Perkins was not one of Silicon Valley's better-known philanthropists, nor did he in fact donate as much of his wealth as others in high-tech. He sometimes even pooh-poohed the trumpeted giving of a Bill Gates or Warren Buffett. But in his own quieter way, he did his share.

Gerd's memory nourished him, but at times it also weighed on him. A year after her death, he was racing *Mariette* off northern Sardinia, in Porto Cervo. Perkins was not religious and at times seemed anti-religion, in part a function of his science background and in part due to his innate skepticism. Nonetheless, early one morning before the first race, he visited a small modern church that had been designed by Joan Miró. He retreated to a pew, alone in his thoughts, and talked to Gerd, as he sometimes had over the prior year. "Are you there?" he asked aloud, his eyes welling up. "I love you. Are you there, somewhere? Can you send me a sign?" Then a ray of sunlight came forth upon his feet, even though the small church windows showed no light. Though he soon discovered that the ceiling had tiny openings in which light could theoretically peer in—this was the empiricist and scientist at work—he chose to believe in another explanation. "I had never experienced a miracle in my life and never have again," he said of that morning in Porto Cervo. "It was a revelation. Gerd was telling me she heard me and that she was there."

There was a limit to how much anyone can dwell on the death of a loved one. You sink into despair or you find ways to move ahead, if only in fits and spurts. Perkins's philanthropy in her name was a step in that direction, a way to remember the good, but there was only so far that it was going to take him emotionally. As important as they'd been before, boats now became an obsession or, in other words, a grand distraction.

In the tales of the sea told over many generations, few are as haunting as the story of the Flying Dutchman—a mariner destined to sail the raging oceans, forever pursuing unattainable love. Be it in Homer's *Odyssey* or Richard Wagner's opera or Sir Walter Scott's poem or Samuel Taylor Coleridge's "Rime of the Ancient Mariner," the themes were the same: loneliness, wandering, everlasting torment. In his darker, introspective moments, Perkins contemplated what immortality might be like. "I cannot imagine it," he brooded. "Eternal consciousness after death—that would be damnation. If you wake up after you die, you'll know you're in hell. Heaven is sleep."

Sitting in the passage cabin of the *Falcon* along the coast of France, before heading for Gibraltar and across the Atlantic, Perkins talked about what would come next. He would naturally try to go faster, looking for mistrals in the Gulf of Lion. By the following summer, he planned on adding a removable one thousand-pound carbon-fiber daggerboard bolted to the bottom of the keel, to make the boat sail closer to the wind. He would consider visiting quiet ports on remote islands in the South Pacific, or major cities around Europe and the Med. He might sail back to Scandinavia or on to Africa or maybe return to Antarctica, which he'd enjoyed so much on his circumnavigation of the world. He might try to break the celebrated speed record from New York to San Francisco around Cape Horn—a record befitting a clipper yacht. The maritime community would be captivated by the attempt, and the media in both cities would behold his great ship.

Yet in whatever he did, and wherever he did it, Perkins would be profoundly alone. When Gerd was dying, she seemed to know her husband better perhaps than he knew himself: she told friends that she hoped he might remarry someday. Perkins did treasure his independence and freedom and solitude—the trappings afforded by money and time. But he had loneliness as well. If there was ambivalence reflected in those two sentiments, it made all the more sense: Perkins's emotions were a mosaic of motivations. Yet either way, he was usually alone on his boats. Even Gerd hadn't made major crossings with him and there were stretches of time when the two of them had been apart. His two children rarely sailed with him. He had no grandchildren. There were few family photographs aboard. His close friends and old colleagues came to visit and stay on the boat, but he seemed just as happy when they departed and he was in splendid isolation again. He delighted in saying he could sail the boat by himself, even if it took a small army to maintain it, clean it, dock it, and prepare a meal.

In making the *Maltese Falcon*, Perkins had done something amazing—creating a boat that might revolutionize sailing. The industry was now referring to the DynaRig as the "Falcon Rig," and Perini Navi was considering building smaller versions of the *Falcon* instead of its traditional ketches. Perkins had conceived of a project that many in the industry thought was too risky, too expensive, and too nuts—and he had seen it through to glorious fruition. His legacy was now not just venture capital, but a new kind of sailing technology. The *Falcon* was a rich man's toy, but it was also a masterpiece of engineering and imagination. Yet his ship had not really come in, after all the years. He remained unsatisfied, unfulfilled, ill at ease. Come the spring of 2008, Perkins had put *Falcon* on the market—for about $235 million (which would yield him a handsome profit of $85 million, not bad even for a venture capitalist). He had learned again that the scheming of the dream outdid the pleasure of the doing. In his pursuit of fair winds and distant shores against an endlessly receding horizon, in seeking to escape his own sense of loss, he had wound up where he'd been in the beginning and where he'd been all along. Ultimately, Tom Perkins was still in search of serenity.

Note on Usage

I have used the pronoun "she" throughout most of this book when referring to the *Maltese Falcon*, except where I thought it affected readability. Reasonable people can surely disagree whether "it" would have been better; many usage manuals recommend "it." But I concluded that this figure of speech dating to the Middle Ages for many inanimate objects—Earth, cities, nations, the seas, as well as ships—was worth following. I thought that was especially so since most folks in the yachting business, including those young enough to have grown up in the age of political correctness, still use the feminine pronoun. Call me a traditionalist, but please never an "it."

Acknowledgments

I grew up sailing with my family on Long Island Sound and along the coast of southern New England. My interest in sailing dates to those times, with the blame resting squarely on my father, whose passion for the water led to much family "togetherness" in small quarters—five kids and a Labrador Retriever named Sheba on a thirty-foot sailboat. My interest in Tom Perkins dates to the late 1990s, when I wrote a bestseller on Silicon Valley, as well as several cover stories for my employer, *Newsweek*, on the culture of the high-tech boom. This book—the biography of a boat and a man—is the result of both my interests in sailing and Perkins.

I would not have been able to write this book without Perkins's cooperation, indulgence, and patience. He is someone who has spent his professional life managing risk, and letting me into his world was a risk beyond his control. I thank him for taking that chance. I'm also grateful to the crew of the *Maltese Falcon*, who graciously accommodated my intrusions onto their boat. Kathy Jewett, Perkins's assistant in San Francisco, is nonpareil—and Perkins is lucky she doesn't fire him. I thank her for answering "just one more question" from me a few hundred times. Danielle Steel, though divorced from Perkins, could not have been more kind with her time and her insight into her former husband.

I have listed at the end of the book all the individuals I interviewed.

But several endured many rounds of fact-checking e-mails or Skype calls: Ken Freivokh, the interior designer; Gerry Dijkstra and Jeroen de Vos, the naval architects, as well as Anneliek van der Linde in their office; Burak Akgül, Cristina Bernardini, Bruce Brakenhoff, Jr., and Sara Gioanola at Perini Navi; Michael Koppstein at Royal Huisman; *Mirabella V* owner Joe Vittoria; journalist David Glenn; and Katrina Arens, Justin Christou, George Gill, and Jed White aboard the *Falcon*. Akgül also acted as a translator for my interviews with Fabio Perini. The photographs for this book, both on the cover and inside, were provided by Emilio Bianchi, Carlo Borlenghi, Bugsy Gedlek, George Gill, Julian Hickman (1blue harbour.com), Franco Pace, Justin Ratcliffe (ratcliffe.justin@gmail.com), Giuliano Sargentini, Jed White, and Tim Wright (photoaction.com). My appreciation to them all.

Thanks also to Peter Carry for translating transcripts from Perkins's trial in France; Cindy Boris at United Airlines and Lee Anderson at Delta Airlines; Nadine Joseph and Neil Goteiner, for assistance in Antibes, France; Fred Shapiro, associate librarian at the Yale Law School; Erica Platt at Morgan Stanley; Laura Sizemore for getting my BlackBerry to work in Istanbul; Anna Tabone in Malta; Ilknur Bayraktar and Elif Özer, for information about the Çiragan Imperial Palace; Owen Mathews, Elif Ozmenek, and Emrah Ulker, for locating and translating Turkish newspaper stories; and editors Jill Bobrow at *ShowBoats International* and Peter Janssen at *Yachting*, for research help. And thanks to Marty Sklar and George Hackett for comments on the manuscript, and saving me from gratuitous puns and countless other errors.

I am fortunate to work for a magazine that allows staff members to pursue outside projects, all the more so when they do it serially. My special thanks to *Newsweek*'s Rick Smith, Mark Whitaker, Jon Meacham, Alexis Gelber, and Kathy Deveny. Barbara Kantrowitz and Daniel McGinn covered for me while I was on the *Falcon*'s maiden voyage, a trip I could not have taken without their help. Dana Gordon, in the *Newsweek* library, found needles in online haystacks that I didn't know existed; Susan McVea and Dave Friedman helped with digital files; Myra Kreiman tracked down archival photographs; and Dan Revitte, Nurit Newman, and Bruce Ramsay attempted to teach me about cover design. I'm glad to count them all as colleagues.

This is my third time being published by HarperCollins. Under difficult circumstances and time pressure, my editor Doug Grad guided this book through. Early on, Henry Ferris was a valuable sounding board as well. Richard Ljoenes did magic with the cover. And Esther Newberg is inestimable among literary agents, despite misguided baseball affections. Thank you, all.

My wife, Audrey Feinberg, read the manuscript and was a constant source of sanity, but her assistance with this project was limitless. All my love to her, and to my two boys, Joshua and Nathaniel.

Sources and Bibliography

The principal sources for this book are interviews I did with the individuals listed below, primarily in 2006 and in a few cases in early 2007. I also had extensive conversations with Tom Perkins in 2004 and 2005, as well as in 1998 for a previous book I wrote, *The Silicon Boys*. My interview with Gene Kleiner was done in 1998, also for that book; he has since died. I used no anonymous sourcing; in one instance, described in the book, I did change the name of a former girlfriend of a crew member, to prevent undue embarrassment and because her name was not central to the story told about her.

Burak Akgül

Prince Albert (Monaco)

Katrina Arens

Nicki Arthur

Timmy Attard

Robert Bell

Sam Bell

Cristina Bernardini

Skip Blair

Jill Bobrow

Kathryn Bosman

Larim Boudjellal

Bruce Brakenhoff, Jr.

John Brickwood

Charles Bushell

Frank Caufield

Karla Christensen

Justin Christou

Jim Clark

Richard Coles

Harriet Davies

Gerry Dijkstra

Viet Dinh

Jerry Di Vecchio

Beverly Dreyfous

John Dreyfous

Robert Eddy

Ken Freivokh

Chris Gartner

George Gill

Sara Gioanola

Baki Gökbayrak

Michael Gooch

Steve Hammond

Mark Healy

Ron Holland

Simon Hutchins

Kathy Jewett

Steve Jobs

Gene Kleiner

Andrew Knight

Michael Koppstein

Rebecca Laing

Vanni Marchini

Nick Maris

Nina Melck

Richard Mondzak

Lachlan Murdoch

Rupert Murdoch

Thys Nikkels

Mete Oktar

Jess Patterson

Fabio Perini

Tom Perkins

Tor Perkins

Roel Pieper

Louie Psihoyos

Peter Quarrie

Giancarlo Ragnetti

Corrado Riccio

Damon Roberts

Bill Sanderson

Patrick Simon

Natalia Singleton

Danielle Steel

Mark Stevens

Allan Stone

Dave Thompson

Franco Torre

Bill Tripp

Christian Truter

Joe Vittoria

Joe Vittoria, Jr.

Jeroen de Vos

Graham Waters

Jed White

Kirsten Wibroe

Knud Wibroe

Liz Windsor

In addition to my interviews and observations, and such original documents as plans for the boat and correspondence between the key individuals, I used a range of magazines and books, as well as one oral history. The University of California, Berkeley, did a series of interviews with Perkins in 2001 on the history of Genentech and his chairmanship of the company from 1976 to 1995; the interviews were for Berkeley's ongoing "regional oral history project" of the West. The transcripts of those interviews, done by Glenn Bugos, were useful in filling in gaps in Perkins's recollections.

I relied on the following publications for information about the yachting industry generally and the *Maltese Falcon* specifically: *Boat International, The Crew Report, Dockwalk, Latitude 38, ShowBoats International, The Yacht Report, Yachting*, and *Yachting World*. Since 2004, the yachting press has been filled with articles on the *Falcon*.

Books were particularly helpful in preparing Chapter II's background on the history of sailing; Brian Lavery's *Ship* was the best. This is the complete list of the books I used for this project:

Chapelle, Howard I. *The History of American Sailing Ships*. New York: W.W. Norton, 1935.

Chapelle, Howard I. *The National Watercraft Collection*. Washington: U.S. Government Printing Office/Smithsonian Institution, 1960.

Dana, Richard Henry, Jr. *Two Years Before the Mast*. New York: Modern Library, 2001 (originally published in 1840).

Jobé, Joseph, ed. *The Great Age of Sail*. Lausanne, Switzerland: Edita S.A. Lausanne, 1967.

Kaplan, David A. *The Silicon Boys*. New York: William Morrow, 1999.

King, Dean. *A Sea of Words: A Lexicon and Companion for Patrick O'Brian's Seafaring Tales*. New York: Henry Holt, 1995.

Lavery, Brian. *Ship: The Epic Story of Maritime Adventure*. New York: DK Publishing, 2004.

Lewis, Michael. *The New New Thing*. New York: W.W. Norton, 1999.

Lubbock, Basil. *The Log of the Cutty Sark*. New York: Charles Lauriat, 1924.

Macaulay, David. *The Way Things Work*. Boston: Houghton Mifflin, 1988.

McKenna, Robert. *The Dictionary of Nautical Literary*. Camden, Maine: McGraw-Hill/International Marine, 2001.

McPhee, Michael. *Sailing*. New York: Fodor's Travel, 1992.

Sherwood, Richard. *Sailboats (2d edition)*. New York: Houghton Mifflin, 1994.

Smyth, W.H. *Chapman: The Sailor's Lexicon*. New York: Hearst Books, 1996.

Somer, Jack. *Athena: A Classic Schooner for Modern Times*. Vollenhove, Netherlands: Stichting Foundation/Royal Huisman Shipyard, 2005.

Villiers, Alan. *The Way of a Ship*. New York: Scribner's, 1953.

Full disclosure: This book was published by HarperCollins, which is owned by News Corp. As I noted in the book, Rupert Murdoch runs that company and Tom Perkins is on its board of directors. Neither one of them had approval rights on the manuscript and neither saw it before publication. Barry Diller, whose boat *Eos* I briefly mention, is on the board of directors of The Washington Post Company, which owns *Newsweek*, which employs me. Diller declined to be interviewed for the book.

Index